The P=NP Question and Gödel's Lost Letter

Richard J. Lipton

The P=NP Question and Gödel's Lost Letter

 Springer

Richard J. Lipton
Georgia Institute of Technology
College of Computing
School of Computer Science
Atlantic Drive 801
30332-0280 Atlanta Georgia
USA
rjl@cc.gatech.edu

ISBN 978-1-4899-9272-7   ISBN 978-1-4419-7155-5 (eBook)
DOI 10.1007/978-1-4419-7155-5
Springer New York Dordrecht Heidelberg London

Printed on acid-free paper

Springer is part of Springer Science+Business Media (www.springer.com)

This is dedicated to my dear wife Judith Norback.

Preface

Does $P \overset{?}{=} NP$.

In just five symbols Dick Karp –in 1972–captured one of the deepest and most important questions of all time. When he first wrote his famous paper, I think it's fair to say he did not know the depth and importance of his question. Now over three decades later, we know $P=NP$ is central to our understanding of computation, it is a very hard problem, and its resolution will have potentially tremendous consequences.

This book is a collection of some of the most popular posts from my blog— **Gödel Lost Letter and P=NP**—which I started in early 2009. The main thrust of the blog, especially when I started, was to explore various aspects of computational complexity around the famous $P=NP$ question. As I published posts I branched out and covered additional material, sometimes a timely event, sometimes a fun idea, sometimes a new result, and sometimes an old result. I have always tried to make the posts readable by a wide audience, and I believe I have succeeded in doing this.

One of the things I think people sometimes forget is that research is done by people. I wanted to make the posts people oriented. I have over the years tried to talk about the "who" as much as the "what." One of my goals—perhaps the main one—is to get you to be able to see behind the curtain and understand how research works. It is a secret agenda of mine to explain what real researchers do when they work on open problems. They make mistakes, they go down dead-ends, they make progress. Students may not know, but even experts sometimes forget, that ideas that we now see as easy were once unknown. This agenda is the main reason that I have made the blog people oriented. Since I have over 30 years of experience working in theory, I know almost everyone that I write about. I think this makes the discussions more interesting, and I hope that you will agree.

I am Dick Lipton, a Professor of Computer Science at Georgia Tech. I have worked in the area of theory of computation since 1973. I find the whole area exciting and want to share some of that excitement with you—I hope that these chapters inform and entertain you. I hope that you not only learn some new ideas, hear some interesting open problems, but also get some of the history of the area. One of the things I think people sometimes forget is that research is done by people. One of my goals—perhaps the main one—is to get you to be able to see behind the curtain and understand how research works.

You should never bet against anything in science at odds of more than about 10-12 to 1 against. Ernest Rutherford, the winner of the Nobel Prize in 1908.

Atlanta Georgia,   *Richard J. Lipton*
                                                                        May 2010

Acknowledgements

I thank all those who supported and continue to support my blog. Thanks to Paul Beame, Rich DeMillo, Lance Fortnow, Bill Gasarch, Timothy Gowers, Gil Kalai, Richard Karp, Michael Mitzenmacher, Terence Tao, Avi Wigderson, Luca Trevisan, and everyone else for their kind support. I also thank the readers of the blog who make it a pleasure to do the work needed to make the it fun.

I am especially thankful for the great support of Subrahmanyam Kalyanasundaram and Ken Regan. Without their tireless support and help there would be no Gödel's Lost Letter blog, and certainly this book would not have been written. I cannot thank them enough.

Finally I thank all the staff at Springer for making this book possible.

Contents

Part I
A Prologue

Chapter 1
A Walk In the Snow

1.1 1956

He use to walk the rolling grounds of the Institute for Advanced Studies with Albert, but Einstein died last spring. Today Kurt Gödel is walking alone through the falling snow, and he is thinking. Gödel is the greatest logician of his time, perhaps of all time, yet he is deeply troubled. He has found a problem he cannot solve—this does not happen often. The problem concerns a simple but fundamental question.

Suddenly he smiles—he has an idea. If he cannot solve the problem, then he will write a letter explaining it to a colleague. He will write a letter to John von Neumann; perhaps John will be able to solve the problem. After all John von Neumann has one of the fastest minds in the world. Gödel pulls his scarf tighter around his neck, gets his bearings in the heavy falling snow, and heads to his office to write his letter.

Unfortunately, von Neumann may have one of the fastest minds of all time, but he is dying from cancer. He receives the letter, but never solves the problem—as far as we know . . . .

1.2 1971

A tall figure walks slowly to the front of the conference room. Steve is a young computer scientist, who is about to change the world. He has independently discovered the the problem that troubled Gödel that snowy day [39]. He gives his talk, and after there is polite applause, as there is for every talk. Yet no one in room appears to be aware the world has changed.

1.3 1972

Dick Karp is in the audience, and he understands what Steve Cook has done. Dick soon writes a paper [76] that is now famous: it states the P=NP question. As more researchers hear about Dick's paper, the importance of his ideas becomes a prime issue. Most agree the ideas raised by Cook and Karp create a new direction for computer science research. Yet, there is no unanimity on how important this direction is. Many wonder—is the direction a "fad," is the direction important, or is the direction a shift that will forever change the world?

Telling how important a new research direction is can be notoriously difficult—especially as events are still unfolding. For example, after Einstein's discovery of special relativity, years passed before physicists understood the importance of his work. Routinely the time between when research is done and when it is awarded a prize, such as the Nobel Prize, is decades. Rarely, is the research community able to immediately judge the importance of radically new research.

However, in this case, one computer scientist, Michael Rabin , does understand how important Cook and Karp ideas are right away. Rabin (who we will later see played a major role in other parts of the P=NP question) "pushes" for a meeting to allow the rapid dissemination of Dick's new ideas. He helps convince Ray Miller and others at the IBM Watson Laboratory to quickly put together a conference at Yorktown—the conference was held in March of 1972. There Karp's presents his work to over two hundred computer scientists—this is a major fraction of the computer scientists at the time.

Ray has the talks from the conference published as a book [106], allowing the whole community access to Karp's paper. Remember, this is years before email and decades before the Internet—thus, publishing the book is a great service to the community.

The reason Dick's paper set off such a storm of excitement is he realized what Cook has done. More exactly, Dick places Steve's work in a central position within computer science. Steve's paper approached the problem from logic—like Gödel had done. The title of Steve's paper is: "The Complexity of Theorem-Proving Procedures." Dick approaches Cook's ideas, instead, from the point of view of practical computer science—this is why Dick's paper galvanizes the community so deeply.

The core of Dick's insight is that he realizes that Cook has proved a key result—we now call the theorem, **Cook's Theorem**. This result is fundamental to all of computer science. The contribution Karp makes is manyfold: he restates Cook's Theorem, he improves the theorem, he names the problem, and most importantly, he shows it has wider consequences than either Gödel or Cook dreamed of.

1.4 Notes

Gödel did write a letter to von Neumann—this is the famous "lost" letter. See the appendix for a translated version. I have created the story of his walk in the snow for dramatic effect. Perhaps Gödel thought of sending a letter in his office, but I like the image and hope you allow me the story. As for the weather, the letter was written in March and it certainly can snow even in April in Princeton—I lived there for years and one of the largest storms ever was in April.

Part II
On the P=NP Question

Chapter 2
Algorithms: Tiny Yet Powerful

Lenore Blum is a computer scientist who is now at Carnegie-Mellon University. She is famous for her work with Steve Smale and Michael Shub on the "The P=NP Question Over The Reals" [18]. This important work shows that there is a version of P=NP for computations over the real numbers. Also she is one of the national leaders in helping get more women more involved in Computer Science. Finally, she is part of the most well known family in all of computer science: her husband, Manny Blum, is a Turing Award winner, and her son, Avrim Blum, is one of best researchers in the world in machine learning.

The work of hers that I am most interested in discussing here is work we did together. We noticed that "algorithms are tiny." Let me explain. Recall that the Gettysburg address is short, only 269 words, and was delivered in just over two minutes. Short things can be very powerful. Here is a note we put together to explain the consequences of the fact that algorithms are tiny:

Algorithms: Tiny yet powerful and we can't live without them.

The promise of information technology is intimately intertwined with advances in algorithms (methods for solving problems). Algorithms are different from software systems in a fundamental way. They are tiny. An algorithm that solves an important problem can be small enough to fit on a single piece of paper using normal sized handwriting. Microsoft's operating system, Vista, is a collection of many programs that in total has close to 100 million lines of code. That's roughly the equivalent of 10,000 hardback books. Vista's code is a trade secret, but even if it wasn't, its not clear a single person could ever understand it all. Algorithms are at the opposite end of the spectrum. A postcard is often enough. As such, algorithms can be shared, understood, analyzed & even proved correct. Algorithms are based on ideas. New algorithms must be based on new ideas, new ways to approach the problem, new thoughts. They often express a clear single thought that changes the way we look at the world. Algorithms are like equations. Albert Einstein's famous equation,

$$E = mc^2$$

is a part of the popular culture. Few non-experts are able to explain the exact meaning of this equation. But, when placed on a T-shirt, most recognize it. This equation is one of the most profound and important discoveries of all time. For better or worse, the equation created the Atomic Age. The plans and blueprints for building a modern atomic reactor may be closer in size to Vista. But the equation that gives us the power to build the atomic reactor has just five symbols. Likewise, it is an algorithm that makes possible the modern cryptographic systems that are critical for e-commerce. Prime numbers form the backbone of these systems. The famous Miller and Rabin algorithm to test whether or not a number is prime contains just 61 words and fits in a box [105, 115]:

> Given input n, an odd integer greater than 1. Write $n-1$ as $(2^s)d$ by factoring powers of 2 from $n-1$. Pick a randomly in the range $[1, n-1]$. If a^d mod $n=1$ or mod $n = -1$ for some d in the range $[0, s-1]$, then return probably prime. Else return composite.

Tiny, powerful algorithms have built the foundations for giant companies: The airline industry, the shipping and logistics industries, and the financial sector all rely on fast optimization, scheduling, and pricing algorithms. Medical imaging devices to detect heart valve damage would not be possible without the Fast Fourier Transform. The list goes on and on. New algorithms will drive yet unknown industries. They will be essential for creating new kinds of computing: petascale, nanoscale, quantum, bio, neuro, distributed, multicore. They speed up today's technology while ushering in tomorrow's. They will be key to facing the challenges arising from the ubiquitous collections of mega sized data sets. In virtually any field where computation is done, the long-run impact of algorithmic advances will, eventually, outrun even the impact or demise of Moore's law. Natural processes such as the regulation of proteins, the dynamics of social networks, and the strategic behavior of companies are all fundamentally algorithmic in nature. Algorithmic theory provides powerful tools, as well as appropriate language and mindset to help understand these new domains. The monumental first mapping of the human genome was accomplished by Gene Myers' whole genome shotgun sequencing algorithm [110]. Exploring the potential of algorithms lies at the heart of the million dollar question: *Does P = NP?*

While research to answer this question may seem esoteric, it lies behind our current ability to have private and secure electronic communications. Our ability to maintain security in the future will hinge on research seeking answers to this question. Advances in our deep understanding of the nature of computation will underlie and propel further advances in virtually every area of information technology.

2.1 A Challenge

Ironically, for secure communication what we desire is the *absence* of an algorithm. There must not exist an efficient algorithm that would enable an eavesdropper to decode our secret messages. Yet, the following simple challenge seems hopeless. Prove there is no algorithm for SAT that runs in polynomial-time and can be written down on a single piece of paper. You can make the paper $8\frac{1}{2}$ by 11, if you wish. The writing has to be normal size. No tricks.

I claim that this challenge is completely hopeless. There is no way that anyone will be able to prove that such an algorithm does not exists. Note, P$\neq$NP says much more–it says that there is no algorithm at all of any length. The algorithm could be a trillion pages in length, or it could be one page. What is so shocking is that the conventional wisdom is confident, yet the one page challenge appears to be hopeless. Does that bother you? I find it shocking that we cannot today rule out that there is a one-page algorithm that solves SAT.

2.2 Open Problems

I have often wondered if there is some hope to prove that restricted algorithms cannot compute SAT. By restricted I mean algorithms that can only have very simple "loop" structure for example. There may be a way to make this precise enough so that we could make some progress. Yet, the full one page challenge is currently out of reach.

2.3 Notes

This appeared as algorithms-tiny-yet-powerful in the blog.

Chapter 3
Is P=NP Well Posed?

Russell Impagliazzo is one of the most original thinkers in complexity theory, and is famous for many wonderful things. One of his papers is on five worlds [70]. The worlds are: *Algorithmica, Heuristica, Pessiland, Minicrypt, and Cryptomania*—see his paper for the definitions.

Let's talk about a different take on P=NP and possible worlds that we live in. I think that the situation is much more complex than Russell's view. As usual, I have a different view of things, which I hope you will find interesting.

Some of you may have been at Russell's invited talk at STOC 2006. It was a great talk, as all his presentations are. But, the coolest part–for me–was that he arrived with transparencies, overheads. The room for the invited talk was very wide and not very deep. So it was critical that his slides appear on two projectors, otherwise part of the audience would be unable to see his slides.

Of course if he had used Powerpoint this would have been automatic, but Russell arrived with plastic slides. The organizers made a last minute slightly-panicked run to Kinko's, where they made identical copies of his colorful slides. Later, as he spoke and turned his slides, the slides on the other projector also flipped–magically. Of course someone had the job of trying to keep this all in synch. It was a great talk, in any event.

Now let's turn first to another famous open problem the Riemann Hypothesis.

3.1 Riemann Hypothesis

The prime number theorem states that $\pi(x)$, the number of primes less than x, is well approximated by the logarithmic integral:

$$\pi(x) = \text{Li}(x) + E(x)$$

where

$$\text{Li}(x) = \int_0^x \frac{dt}{\log(t)}$$

and $E(x) = o(x/\log x)$ is the error term.

The Riemann Hypothesis (RH) implies that the error term, $E(x)$, is order $x^{1/2}$—ignoring logarithmic terms. But, the RH is "binary" in the sense that having even *one* zero that is not on the critical line destroys this bound. More precisely, if there is a zero at $\sigma + it$ with $\sigma > 1/2$, then the error term in the prime number theorem is at least x^{σ}–again ignoring some logarithmic factors. I thank Andy Odlyzko for pointing out this tight relationship between a non-critical zero and the error term.

This is a relatively simple behavior: one bad apple and the whole distribution of the prime numbers changes. Famously, Enrico Bombieri has said that "The failure of the Riemann hypothesis would create havoc in the distribution of prime numbers."

3.2 P=NP is Different

I claim that our problem P=NP is different from the RH, since there is no "one bad apple" behavior. **Consider first what a proof that P=NP could look like.**

• There might be an algorithm for SAT that runs in a reasonable time bound. This would be an immense result, and would change the world. Crypto-systems would all be destroyed, whole parts of theory would have to radically change. This is why many people believe that P=NP is impossible–at least that's my impression–they say it would be too good to be true. I agree that it would be an amazing situation, but it is only one of a multitude of possibilities.

• There might be an algorithm for SAT that is polynomial but with a very high exponent, for example, $O(n^{10})$. This would win the million Clay dollars, but would have a more philosophical impact. Crypto-researchers would worry that eventually the algorithm could be made practical, but it would not have any immediate practical impact. Research papers would have to do a song and dance: "we study an approximation to the following NP-hard problem, that is in P of course thanks to the famous result of X, but we feel our results are still interesting because ..."

• There might be an algorithm that is polynomial in a very weak sense. What if its running time is

$$n^{2^{2^{2^{2^{100}}}}}.$$

That would be great, I think you still get the money, but crypto people could still sleep at night, without any sleep aids.

• There might exist an algorithm that runs in n^c time for *some* constant c that is not even known. Results like this are quite possible, if the proof of the algorithm's correctness used some indirect argument. Again you still get the cash–I believe.

Now consider what a proof that P is not equal to NP could look like.

• SAT might require exponential time. The best algorithm might run in time 1.2^n, for example. This type of behavior is what many believe. There even is a "formal" conjecture that the best algorithm for 3-SAT will have this type of running time. Note, even here one needs to be careful: what if the running time is 1.0000001^n? Alan Perlis has said:

> For every polynomial algorithm you have, I have an exponential algorithm that I would rather run.

• SAT might have wild complexity. Perhaps for some lengths n the best running is exponential, but infinitely often the best time is polynomial in n. This could happen—see Chp. 9.

• SAT might be barely super polynomial: what if the lower bound is

$$n^{\log\log\log n}.$$

Does this really mean anything? Yet it would technically resolve the P=NP question, would be an immense achievement, would win the Clay Prize, and probably many other prizes too. But does it really imply that SAT is "intractable?"

• SAT might have no deterministic polynomial time algorithm, but there could be randomized algorithms ...

3.3 The Riemann Hypothesis Again

You might be thinking what if the RH is true, could it have the same complex behavior that P=NP can? In particular, could the RH be true, but with a crazy error term:

$$2^{2^{2^{100}}} x^{\frac{1}{2}} \log(x)?$$

Isn't this the same issue that I raised above? The answer is that this *cannot* happen for the RH: The following is a theorem:

Theorem 3.1. *If RH, then for all $x \geq 2657$,*

$$|\pi(x) - Li(x)| < \frac{1}{8\pi}\sqrt{x}\log(x).$$

This is due to Lowell Schoenfeld, who actually proved that RH is equivalent to his inequality [120].

3.4 Is P=NP Ill Posed?

I love the P=NP question, but it occurs to me that in a sense the question is "ill posed." David Hilbert said once:

If I were to awaken after having slept for a thousand years, my first question would be: Has the Riemann hypothesis been proven?

After sleeping for a thousand years knowing if P has been proved to be not equal to NP, would not be nearly as meaningful as knowing that RH has been proved.

To make my point consider three players: Paula, Ted, and George. Paula is a practical minded computer scientist, Ted is a theory minded computer scientist, and George is the best gossip in the department.

There are two scenes. In the first P=NP and in the second P≠NP. Note FOCS is one of the premier theory conferences: Foundations of Computer Science.

George barges into Paula's office without knocking. She is about to complain that she is in a meeting with Ted, when George says, "Ted did you hear that P=NP has been proved! I just got off the phone with Lance and the proof is correct."

Paula sees Ted jump up at this and he says, "are you kidding?" George is out of breath, but smiling. It's no joke. "Unbelievable, unbelievable," is all that Ted can manage.

Paula has never seen George this excited, "no it's really true. The FOCS program committee accepted the paper, even though she did not submit it. I guess that's a first, she just missed the deadline or something."

Paula sits behind a large desk that is covered in books on circuit design and VLSI, says "this is great. Finally, you theory guys have done something that I can use. I can't wait to get my team implementing the algorithm." She sees Ted sit down, he is turning pale and mutters to himself, "Dammit, this wipes out Fred's thesis. I told that idiot to finish it last spring."

Paula ignores him, "this will change VLSI design and testing and–"

George interrupts her, "well not quite Paula. I hear that the running time of her algorithm is bad."

"What? How bad is bad?"

"I think its running time is $n^{2^{2^{100}}}$, although I forget if there is a logarithmic factor too. But, P=NP!"

George barges into Paula's office without knocking. She is about to complain that she is in a meeting with Ted, when George says, "Ted did you hear that P≠NP has been proved! I just got off the phone with Lance and the proof is correct."

Paula sees Ted jump up at this and he says, "are you kidding?" George is out of breath, but smiling. It's no joke. "Unbelievable, unbelievable," is all that Ted can manage.

Paula has never seen George this excited, "no it's really true. The STOC program committee accepted the paper, even though she did not submit it. I guess that's a first, she just missed the deadline or something."

Paula who is listening from behind a large desk that is covered in books on circuit design and VLSI, idly checks her Seiko. It's almost twenty to 12, she needs to leave for her lunch meeting soon.

Ted says, "I knew it, that crazy blogger–what's his name?–always talking about P could equal NP. Well he was wrong. What an idiot."

George says, "yeah."

"So how does the proof go?"

"It's a circuit lower bound, she proves SAT requires at least $n^{\log\log\log n}$ size circuits. Avoids the usual roadblocks by using that new combinatorial lemma of Terry. Isn't this wonderful."

Paula who a moment ago was hardly listening says, "Did I hear you right? That's not a very big bound. Let's see if n is a million, let me see, that's not even a quadratic lower bound. Were those log's base two?"

"Of course they're base two, log's are always base two. But it's a super polynomial lower bound, P$\neq$NP!"

My apologies to all fiction writers that have ever lived.

3.5 Open Problems

My point is simple: There is a huge variance in the potential answers to the question:

Does P equal NP?

It's not simple. Equal, not equal, Equal, not equal. There are an infinite number of possibilities. I think that this is what makes me think that we need to work hard on this problem.

A class of open problems is to prove something like Schoenfeld's theorem for P=NP. Can we show that some of the above pathological behaviors cannot occur? Or can we show that one pathological behavior implies another? Or precludes another?

3.6 Notes

This appeared as is-pnp-an-ill-posed-problem in the blog. It generated a number of interesting comments. Gil Kalai pointed out he has discussed similar ideas on Russells 'multiverse' with a semi-popular exposition on his blog. Drew Byrne summed it up, "Ah, to equal P or not to equal NP, is that the question?"

Chapter 4
What Would You Bet?

Pierre-Simon Laplace, was a famous mathematician from the 18th century, who is known for many things. How about having a transform named after you—the Laplace transform—pretty cool. He is also famous for posing the scientific question, "what is the probability the sun will rise tomorrow?"

Today I will discuss our version of his question: "What is the probability that $P \neq NP$?" Okay, it not really the same as his question, but it is an interesting question.

4.1 You Bet Your Life?

I raised the issue of what is the right bet about $P=NP$ on my blog once. There was a great deal of discussion, which was extremely interesting. I thank everyone who was kind enough to make a comment.

The central discussion, however, I think moved over to a discussion of beliefs vs. betting. This is a cool topic, but not exactly what I had in mind. I am not really interested in starting a betting pool on $P=NP$. I did want to make this point:

The theory field has placed a bet that $P=NP$ is impossible.

Here is how the field makes that bet every day:

• **Money:** Try sending a proposal to NSF, for example, and state that you wish to work on proving $P=NP$. Or even a weaker goal, that you wish to look at the possibility that factoring is in P. The chances that you will get funded are close to zero.

I tried an experiment a few years back: I wrote a reasonable—in my opinion— proposal on attacks on factoring. I am still thinking about them, and have posted on one of my ideas. The proposal got 0 dollars. Perhaps it was poorly written, or perhaps the funds were tight that year, or perhaps my ideas were stupid, all that could have been true. But, the written reviews made it clear that I could not be funded without a "partial result." If I had a partial result, then I probably would not need NSF funding. You either can factor or not.

The point of this is the field has made a huge bet that P is not equal to NP through the types of proposals that are funded. If P=NP is really closer to 10 : 1, then I would argue that more dollars should flow into different projects.

• **Eye Balls:** Try sending a paper to a conference that goes against the conventional wisdom that P=NP is impossible—I think you will get nowhere. The same for journals or other places that control what we get to see. Again I think that the field has made a bet; the bet is that such papers are uninteresting. For example, would a paper that assumed that factoring was easy and then derived consequences of this assumption get attention? I think that it would not.

• **Mind Share:** Try getting people to think out-of-the-box about P=NP, it's hard. I started my blog mostly to do just that. I am happy that people seem to enjoy the blog; however, I would like people to think about P=NP in a serious way.

Princeton/IAS held a meeting on Barriers in Computational Complexity. The organizers and the speakers were some of the best theorists in the world. I really wish I could have been there to hear such top researchers speak on their research. However, my understanding is that the tone of the meeting was:

How can we finally prove what we already know is true.

My student Shiva Kintali, who was at the meeting, had two cool comments to report. The first was,

One speaker mentioned that once a Nobel prize winner in physics was asked "if you can ask one bit of information from an all-powerful alien what would you ask ?" He said "I would ask if P=NP." At this point many people in the audience said "We all *know* that bit, all we need is a proof." This was surprising to me (Shiva).

The second was that in almost every panel discussion, everybody unanimously said "I believe P≠NP, since a world with P=NP would be hard to imagine." The most common reason they believed P=NP is impossible, Shiva says, is that in this world, creativity would be automated.

That is a pretty strong bet that P=NP is impossible. They may be right, but let's at least agree that they are making a strong bet.

• **National Security:** Try getting security people to worry about a classic factoring algorithm. Good luck. I think there are national security issues here—by "national" I mean more than the US. Any country's safety and economy is potentially at risk.

What if the "wrong people" figure out how to factor, or how to compute discrete logarithm, or in general how to solve hard problems, what will the consequences be? I think at a minimum there should be plans on how to react quickly if the basic assumptions that modern cryptography are based on are destroyed.

This is another implicit bet. If you think that it is impossible that P=NP, then you will probably not plan for the failure of your crypto assumptions. Is this not a bet of a million to one? I say this because the consequences of a failure could cost hundreds of billions, perhaps trillions—the latter if national security is compromised.

4.2 Open Problems

I wanted to point out that we do bet every day on P=NP, whether we realize it or not. I think that given the potential for many outcomes to the P=NP question, we should be more open minded about the computational world. As I have pointed, in Chp. 3, out P=NP could be true and have nothing to do with "automating creativity," or having a practical algorithm. But it still could true.

4.3 Notes

This appeared as theory-has-bet-on-pnp in the blog. This post caused a great deal of discussion. I took an extreme position, and many were less pessimistic than I am. But, again thanks to all who added their comments. Thanks to Bhupinder Anand, Dave Bacon, Howard Barnum, Paul Beame, Stefan Ciobaca, Steve Flammia, Timothy Gowers, Paul Homer, Jonathan Katz, Pascal Koiran, Daniel Lemire, Stefan Lucks, Michael Mitzenmacher, Martin Schwarz, John Sidles, Janos Simon, Giora Slutzki, Aaron Sterling, Tomasz Wegrzanowski, and others. Thanks again.

Chapter 5
What Happens When P=NP Is Resolved?

Ed Nelson is a senior faculty member of the great Princeton Mathematics Department. He is well known for his work in a number of areas, including logic and probability theory. Nelson, in 1950, created a *beautiful problem* about coloring the plane. Not coloring a graph, coloring the whole Euclidean plane.

> *What is the fewest number of colors sufficient for coloring the plane so that no two points with the same color are a unit distance apart?*

I believe this is still open.

Let's talk about what it might be like to have someone resolve P=NP. Back in the mid 1990's Nelson had an approach to proving P≠NP that looked promising, and created quite a stir at Princeton. In Ed's case he thought for a while that he had a proof that P≠NP. The proof was based on some insights that he had from his previous work in logic. While the proof did not work, it did teach us a few things about what might happen if someone really solved the problem.

5.1 The Claim

One day, at Princeton, Andy Yao grabbed me and told me that Ed Nelson had a draft of a proof that P≠NP. I was excited of course, and asked Andy if he had checked the proof yet. Andy said that the paper was still a very rough draft, it had missing sections, and since he had just started to look at it, would I like to help him figure out what was up? I immediately said that I would. Andy gave me a copy of the paper and we started to look at it immediately.

It was clear—instantly—that the paper was from a professional mathematician, although there were some oddities. For example, there were some standard results from complexity theory that Ed proved from scratch, rather than just cite them. This is usually a "red-flag" that the author is not well versed in theory, yet to me it did not rule out that there might be something important in his paper.

Andy and I decided that the paper was sufficiently sketchy that we would have to talk to Ed directly about the details. Andy called Ed and we setup a meeting that was to take place in the computer science building in two days. The goal of the meeting was simple: go over his proof section by section to figure out whether or not it was correct.

5.2 The Meeting

What started as a private meeting of Ed, Andy, and myself quickly snow-balled into a mini-conference. Somehow word got out that *the problem* had been solved. When we met with Ed we had to use a large classroom to hold the number of people who hoped to see history being made.

The meeting was "run" by Avi Wigderson, Andy, and myself. Of course Ed was there as an "oracle" that we could call on to ask questions about his proof. But, the plan was that we would try to go over the proof and only use Ed when we got stuck somewhere.

Ed's proof was based on a novel definition of a family of computations classes,

$$\mathscr{C}_1 \subseteq \cdots \subseteq \mathscr{C}_m \subseteq \ldots$$

Each class really was defined by a restricted class of logical formulas. The base class $\mathscr{C}_1$ was very weak and had only simple operations like addition, multiplication, and equality. Then, each class $\mathscr{C}_{i+1}$ was the last class with a type of restricted quantifier added. Finally, he needed that each class was computable in linear time, and that each class had a certain technical property X that was the key to his whole proof.

If all this worked, then he could prove a separation theorem, that would imply that P$\neq$NP. This is a rough outline, as best we could figure out at the time.

The main structure was fine: the classes nested just as claimed, and if all was true the last step of proving P$\neq$NP would follow. The base class $\mathscr{C}_1$ could be shown to have property X by a nice argument.

The first rub, curiously, was the claim that the base class could be computed in linear time. Now multiplication is computable in nearly linear time, but Ed's proof needed that the time was exactly linear. Nothing higher would work. Avi and Andy tried some different ideas at the board, and we asked Ed. He "forgot" how this step worked–we were stuck. This was scary that we were stuck at the first base case.

Then, I realized a trick. By a simple application of the Chinese Remainder Theorem (CRT) we could check that $x \times y = z$, which was all that was needed. This is when—I recall—Avi said that the CRT was the only theorem that I knew. He says that he never said this. But, in any event the CRT theorem did allow us to get past the first case—we were unstuck. The class $\mathscr{C}_1$ was indeed computable in deterministic linear time.

We then turned to the inductive step. It was here that we got stuck again, since we had been at the proof for a couple of hours we all agreed to adjourn for the weekend. We planned on getting together again the next Monday.

5.3 The Outcome

Since the meeting ended without any conclusion, we all left confused. We were starting to understand what Ed was trying to do, but we were all still confused. At least I was.

I decided that since it was a Friday I would try to spend the weekend either finding the bug or figuring out if the proof was correct. I sat down that Friday night and began to think about what was happening, when the phone rang. It was Michael Rabin calling from Harvard. Michael had heard the rumor, and tried to reach Avi to see what was up. Failing that he called me.

Michael said that he knew Ed Nelson well, and that Ed had done very solid research over the years in a variety of areas. Michael had not seen or even heard any details of the proof, but was excited that it might actually be correct. I told Michael that I was still trying to understand Ed's ideas and I was still confused. We ended the conversation saying that I would get back to Rabin when I knew what was up.

In a few hours I figured out the problem. The proof of the first inductive step worked: the class $\mathscr{C}_2$ had the properties that Ed needed. But, the next step of the induction failed. Once the class became powerful enough his property X no longer held. What had thrown Ed off and confused us was that the induction did not fail immediately. I re-checked my observations and soon I was sure that the proof was wrong, and moreover it could not be fixed.

I reached Avi who also by then had also found the same error, and we then called Michael. The next week Andy and I spoke to Ed and explained why his proof could not work. He listened carefully, and after a few days of thinking acknowledged back to us that his proof did not work. Oh well. It would have been cool to see a solution.

I thought the proof was dead, the claim was withdrawn, and P=NP was back as a completely open problem. But, I was too premature.

5.4 The Proof Returns

A year later Anastasios Viglas and I proved a simple but nice result about the complexity of SAT [92]—see Chp. 14. A few days later I was visiting Telcordia for the day, and I told someone about Viglas's nice theorem, which was to be a part of his thesis. They said that sounded interesting, but they had heard that Ed Nelson had resolved the whole question—he had proved P≠NP. I said I was aware of his earlier proof, but that had fallen apart. They said that they thought this was a whole new proof, and added the interesting detail that the math department had just had a big party to congratulate Ed on his work.

My first concern was pretty narrow: I worried that Viglas' theorem would be wiped out if Ed was correct, his thesis would be set back, and in general he would be in trouble. Perhaps this is a natural protective reaction advisors have toward their students. Of course, a proof that P$\neq$NP would have much greater consequences then just wiping out Viglas' thesis.

I had to find out what the true story was—was there any chance that Ed had actually fixed up his proof? I doubted it, but I needed to know. Since I was at Telcordia, I could not just run over and see my friends in theory or over at math. I had to try and reach people by phone or email.

Just my luck everyone was out that afternoon. No one answered the phone, my emails were not returned, and I had no luck finding out anything.

I finally got the scoop the next day, when I was back in Princeton. Ed's party was a celebration for something else. He did not have any new proof, and so our small result was safe.

5.5 Open Problems

The main problem is to resolve P=NP. But, I think this episode raises some interesting questions. We spent a lot of time and energy debugging a proof that was wrong. Rumors started to fly—a few less doubts and things would have completely spiraled out of control. Newspapers would no doubt have been the next to jump in, if we had not found the serious bug in the inductive proof as quickly as we did.

How should we handle claimed proofs from professional serious researchers?

5.6 Notes

This appeared as what-will-happen-when-pnp-is-proved in the blog.

Chapter 6
NP Too Big or P Too Small?

Leonid Levin is a computer scientist who independently proved essentially the same theorem as Steve Cook did. They both did this in 1971 [90, 39]. This was during the height of the cold war and the information flow between Russia and the west was not very smooth. So it took a while for the theory community to understand what Levin had done. Levin is also famous, in my opinion, because his advisor was perhaps one of the most dominant mathematicians of the 20th century: Andrey Kolmogorov. Kolmogorov solved long standing open problems as well as created entire new fields of mathematics. This is a powerful and unique combination.

6.1 Too Big or Too Small?

One approach to P=NP is to show that SAT requires super-polynomial sized circuits. By a beautiful theorem of John Savage every problem in P has polynomial size circuits [118]. John Savage Therefore, if SAT required super-polynomial sized circuits, then this would yield a contradiction. Thus, a lower bound on circuits can yield a proof that P is not equal to NP.

This is the approach that many have tried to follow: somehow show that SAT's circuit complexity is large. So far the progress in this direction is zero. I believe that no one can even prove that SAT requires at least a circuit of size $1.0001n$ where n is the number of bits in the encoding of the problem. Note, there are some artificial problems that have been shown to require at least $3.5n$ size circuits. However, they are not SAT.

There is an alternative approach to P=NP. Recall, the story of the "Goldilocks and the Three Bears." Recall she finds that one bed is too small and another is too big. Finally, one is just right. Well there is a corresponding approach to our favorite problem.

Theorem 6.1. *Suppose that all problems in P have circuit complexity at most $O(n^c)$ where c is a universal constant. Then, P$\neq$NP.*

I have known this theorem for a long time, but do not know who first proved it. The proof of the theorem is the following: If P$\neq$NP, then we are done. So assume that P=NP. Now by a pretty result of Ravi Kannan NP must have a problem with circuit complexity at least $\Omega(n^{c+1})$ [73]. But, for n large enough this is a contradiction. This completes the proof of the theorem.

The cool part of this argument, I believe, is that it shows there are two ways that P and NP could be different. One is that NP is "too big", i.e. it has no fixed sized circuits. Of course this is where the conventional wisdom is betting. Most would agree that SAT must have big circuits. Yet there is no evidence to this. None. In my opinion it is completely possible that this could be false.

The other possibility is that P is "too small." In this case P has fixed sized circuits and so it is much smaller than NP. Of course, conventional wisdom does not believe this is possible. I think it could be the case. Circuits are very powerful. They can do strange things that algorithms cannot. A classic example of this is the famous result of Len Adleman on de-randomization of circuits [2]. So I think it is possible that this could be the way that P=NP gets resolved. But who knows.

6.2 Open Problems

While there is currently no evidence whether NP is too "big" or P is too "small," Levin surprised me one day while we chatted about P=NP at a conference. He remarked that his advisor, Kolmogorov, believed that P had circuits of size $O(n)$, i.e. linear. (Of course Kolmogorov stated his ideas in a different way since P for example was not defined explicitly back then.) Linear is as small as they can be, of course. Levin insists that the great man had really thought about this. While Kolmogorov had no concrete results he was convinced that linear was the right answer. I have only Leonid's word for this, but it certainly made an impression on me.

Clearly, the obvious open problems here are several. One possible tractable one is to prove that SAT itself has circuit complexity at least $2n$. While this has no major consequence, it seems to be open and might shed some light on SAT. Another set of questions concerns trying to improve Savage's theorem. His famous theorem proves if a Turing runs in time $T(n)$ then for all n there is a circuit of size $O(T(n)^2)$ that solves the problem on inputs of length n. This was improved long ago to $O(T(n)\log T(n))$. Can we prove anything better? If Kolmogorov is right, then it should be possible to vastly improve this bound.

6.3 Notes

This appeared as is-np-too-big-or-p-too-small in the blog.

Chapter 7
How To Solve **P=NP**?

Landon Clay is the Boston businessman who created the Clay Institute and the associated Millennium Prize Problems, one of which is P=NP. The solver(s) of P=NP will get, after certain rules are fulfilled, a cool one million dollars. That is a lot of money, but somehow juxtaposed next to current bank bailout and other loans measured in the trillions of dollars, it does seem a bit less impressive. But, unlike Grigori Perelman, I would take the money, for sure.

In any event, let's talk about how I would approach the P=NP question, if I were forced to work on it. I think of the quote from Paul Erdős on Ramsey numbers:

> He asks us to imagine an alien force, vastly more powerful than us, landing on Earth and demanding the value of $R(5,5)$ or they will destroy our planet. In that case, he claims, we should marshal all our computers and all our mathematicians and attempt to find the value. But suppose, instead, that they ask for $R(6,6)$. In that case, he believes, we should attempt to destroy the aliens.

My version is suppose that the alien force asks us to prove that P=NP. I would say get our best theorists and mathematicians and look for an algorithm. If on the other hand, they ask for a proof that P$\neq$NP, I would agree with Erdős and suggest we attempt to destroy them.

You know why I believe this. Even if the conventional wisdom is that P=NP is false, I think we have a better chance of finding an algorithm, then a lower bound proof. I guess also if the aliens are "fair," by asking for a proof that P=NP, one could conclude that it was true. To ask for a proof otherwise would make the aliens pretty mean. If they were that evil, they might destroy us whether we solved P=NP or not.

There is a history of people asking for solutions to problems that they know are unsolvable. A famous example is the "15 puzzle," which you probably have played with one time or other. The goal is to slide the 15 squares around a 4×4 square until they are in order. I was always told that the great puzzle maker of the late 1880's, Sam Loyd, created the puzzle. The current understanding is that he did not, and the puzzle was created by Noyes Chapman, the Postmaster of Canastota, New York. My sister-in-law lives in Canastota, but that has nothing to do with the puzzle. The designer of the puzzle, whoever it was, was a bit "evil." There is a parity argument showing that the initial arrangement cannot be moved to the required final arrangement. The puzzle is unsolvable. During the craze when the puzzle was popular, there were people who even thought they had solved the puzzle, but could not remember how.

7.1 Role of Prizes

In the earlier history of mathematics, prizes for solving problems played a larger role than today. I talk about Henri Poincaré prize winning paper on the three body problem in Chp. 14. Erdős often offered money for the solution to open questions. He sometimes made the prize money quite asymmetric: a counter-example was worth $50 and proof was worth $500. I recall that one lucky solver won his prize of $1,000 for a very hard problem of Erdős and was asked was it worth it: he answered that he calculated that he made about 50 cents per hour.

Ron Graham used to manage Erdős money, including paying people who solved his open problems. Ron told me that at first most people kept the checks, probably framing them–this saved Ron the money. Then, people got color copier machines, so they could make a perfect copy, frame that, and then cash the check. Ron said that was okay, but what really got him was when some people started to cash the checks, and then asked for the cancelled check back.

A final story about prizes. Jean-Paul Sartre won the Nobel prize for literature in 1964. Apparently he refused the Nobel prize, saying:

> "It is not the same thing if I sign Jean-Paul Sartre or if I sign Jean-Paul Sartre, Nobel Prize winner. A writer must refuse to allow himself to be transformed into an institution, even if it takes place in the most honorable form."

However, later he wrote a letter to the Nobel committee asking for just the prize money; he was politely turned down.

7.2 My Approach

Okay, the aliens have thrown down the challenge to find an algorithm for SAT. He is my best guess today for a reasonable approach. I would first use the powerful results from PCP that show that I "only" need to find an algorithm for 3-SAT that beats the trivial bound of satisfying 7/8 of the clauses. The exact theorem that we are using is:

Theorem 7.1. *Let* $\varepsilon > 0$. *Supposed there is a polynomial-time algorithm that on all satisfiable cases of* 3-*SAT outputs an assignment that satisfies at least* $7/8 + \varepsilon$ *clauses. Then* P=NP.

Let $\mathscr{C}$ be a set of 3-clauses that are satisfiable. I will break the approach into two cases.

• In this case I will assume that the clauses $\mathscr{C}$ are "random." This means that the clauses have enough "randomness", in some sense, that the known algorithms that work well on random SAT can work on these clauses. I have no exact idea how to make this precise, but I hope you see the idea.

• In this case I will assume that the clauses are not "random." This means that there is some inherent structure to the clauses. I will hope that I can exploit this structure to show that there is a fast method for solving the SAT problem on structured clauses.

Such randomness "vs." structure is the idea that was used so brilliantly by Ben Green and Terence Tao in the recent famous proof on primes in arithmetic progressions [60].

7.3 Open Problems

The obvious question is can we make any progress on P=NP using this or some other approach. What I like about the approach is two things. First, it uses one of the deepest results known in computational complexity–the PCP theorem. Second, it is looking for an algorithm, which we are good at. Just about every breakthrough in theory can be viewed as a clever new algorithm. This even holds for lower bounds. For example, the bounds on constant depth circuits are really based on clever algorithms for approximating certain kinds of circuits by low degree polynomials.

7.4 Notes

This appeared as how-to-solve-pnp in the blog. Both Timothy Gowers and Timothy Chow had very interesting comments on whether this approach to the P=NP question is reasonable.

Chapter 8
Why Believe P Not Equal To NP?

David Letterman is not a theorist, but is an entertainer. I do not watch his show—on too late—but I love to read a recap of his jokes in the Sunday New York Times. He is of course famous for his "Top Ten Lists."

I plan to follow Letterman and try to list the top reasons why people believe that P≠NP. I am not planning on being funny—I have tried to faithfully capture some of reasons that people give for believing this.

The list I have put together has many reasons: they range from theorems, to social believes, to analogies to other mathematical areas, to intuition. I have tried to be inclusive—if you have additional reasons *please* let me know.

One thing that is missing from my list is experimental evidence. A list for the Riemann Hypothesis might include the vast amount of numeric computation that has been done. Andrew Odlyzko has done extensive computation on checking that the zeroes of the zeta function lie on the half-line—no surprise yet [113]. Unfortunately, we do not seem to have any computational evidence that we can collect for the P=NP question. It would be cool if that were possible.

It may be interesting to note, that numerical evidence in number theory, while often very useful, has on occasion been wrong. One famous example, concerns the sign of the error term in the Prime Number Theorem: It was noticed that $\pi(x) - \text{Li}(x)$ is never positive for even fairly large x; hence, the conjecture arose that

$$\pi(x) > \text{Li}(x)$$

is impossible. Here $\pi(x)$ are the number of primes less than x, and $\text{Li}(x)$ is equal to

$$\int_0^x \frac{dt}{\ln t}.$$

This would have been an interesting fact, but John Littlewood proved a wonderful theorem that showed that not only was $\pi(x) > \text{Li}(x)$ possible, but that $\pi(x) - \text{Li}(x)$ took on both positive and negative signs infinitely often [63]. At first the bounds were huge for the first x so that $\pi(x) > \text{Li}(x)$, on the order of the Skewes' Number $10^{10^{34}}$. Today I believe it is still open to find the first x.

8.1 Belief vs Proof

For scientists, especially theorists and mathematicians, what does it mean to believe
X when there is no proof that X is true? I think that there are plausible reasons that
you may or may not believe something like P$\neq$NP without any proof—I plan to list
some of those in a moment. But at some level beliefs are unique to each individual,
perhaps the sum total of all their personal experiences.

However, I think that listing potential reasons for believing an unproved statement X can help guide a field. As I posted earlier, if the theory field is really convinced that P=NP is impossible, then this conviction will manifest itself in many
ways: what is funded, what is published, and what is thought about.

While this is a reasonable use of a belief, there are some dangers. The main
danger is that if the belief is wrong, then there can be negative consequences. If
most believe that some unproved X is true, then:

- Few may work on disproving X, which could potentially delay the final resolution
  of the problem.
- The field may build an edifice that is deep and rich based on X, but the foundation
  is faulty. If many theorems of the form "If X, then Y" are proved, then when X
  is finally disproved a great deal of hard work will become vacuous.
- Finally, even if X is true, by thinking about both X and its negation, the field may
  discover other nuggets that would otherwise have been missed.

8.2 Three Choices, Not Two

Before I present the list I want to be clear that in my mind there are three choices—
actually there are many more as I have pointed out before, but let's stick to three:

1. P$\equiv$ NP: Let this mean that P is really equal to NP, that there is a practical
   polynomial time algorithm for SAT. Note, even this does not imply there are
   practical algorithms for all of NP, since the running time depends on the cost of
   the reduction to SAT.
2. P$\asymp$NP: Let this mean that there is a polynomial time algorithm for SAT, but
   the exponent is huge. Thus, as a theory result, P does equal NP, but there is no
   practical algorithm for SAT.
3. P$\neq$NP: This will mean that there is no polynomial time algorithm for any NP-
   complete problem.

As I pointed out in. Chp. 3 there are even other possibilities. But let's keep things
manageable and only use these three for now.

8.3 A "Why P≡ NP is Impossible" List

Here are some of the reasons that I have heard people voice for why they believe that P≡ NP is impossible:

1. P ≡ NP would be too good to be true. Some have said that it would end human creativity.
2. P≡ NP would imply that thousands of NP-complete problems are all in P. Since they are "different" problems, the argument is that it is hard to believe there is a universal method to solve all of them.
3. P≡ NP would put an end to modern cryptography, which is based on even stronger assumptions than P≠NP.
4. P≡ NP would go against results that are known already for finite automata and pushdown automata. In the first case nondeterminism can help decrease the state size exponentially; in the latter case, nondeterminism adds power to pushdown automata.
5. P≡ NP would go against known results from recursion theory. In particular, the analog of the polynomial time hierarchy is the arithmetic hierarchy, which is known to be strict.
6. P≡ NP would go against known results from set theory. In particular, the analog of the polynomial time hierarchy is the strict hierarchy of descriptive set theory. Mike Sipser did some pretty work on this connection.
7. P≡ NP would make $NTIME(n) \subseteq DTIME(n^c)$ for some c. The brilliant work of Wolfgang Paul, Nick Pippenger, Endre Szemerédi and William Trotter shows that $c > 1$, which is a far distance from proving that there is no c, but it at least shows that $c = 1$ is impossible [43]. That is nondeterminism does *help* for Turing Machines.
8. P≡ NP would show that there is a general search method that is much better than brute force. For SAT, for example, the best known results are that 3-SAT can be solved in exponential time c^n. Rainer Schuler, Uwe Schöning, Osamu Watanabe gave a beautiful algorithm with a $(4/3)^n$ upper bound, while the best is around $(1.3)^n$ [65].
9. P≡ NP would imply that tautologies could be checked in polynomial time. There are beautiful results for specific logics, such as resolution, that prove exponential lower bounds on the proof length of tautologies. There are also lower bounds on even stronger logics: polynomial calculus, cutting planes, and constant-depth Frege systems.
10. P≡ NP would imply that thousands of NP-complete problems are all in P. These problems have been attacked by very different people, with different backgrounds and tools. The fact that no one has found a polynomial time algorithm for any of the problems is strong evidence that there is no such algorithm.

8.4 A Comment On The List

I do want to point out that the some of the list would change if we replace $P \equiv NP$ by $P \asymp NP$. For example:

- $P \asymp NP$ would be *not* too good to be true, since no practical use could be made of the result.
- $P \asymp NP$ would not put an end to modern cryptography—crypto-systems would still be safe.

8.5 Open Problems

What are your major reasons for believing $P \equiv NP$ is impossible? Have I missed some interesting reasons?

Some people would include observations like: no one has found an algorithm for $P \equiv NP$, so it is unlikely to exist. I find this pretty weak—science problems can be open for a long time, and further why is this biased towards $P \equiv NP$ and not other possibilities. Scott Aaronson's own take on a list includes reasons like this.

Another reason mentioned sometimes is $P \equiv NP$ would help solve all problems. Note, this is not true for $P \asymp NP$, and even if $P \equiv NP$ the encoding of your question could be impractical. For example, consider the encoding of this question into SAT:

Is there a formal proof of the Riemann Hypothesis in Peano Arithmetic of length at most 10^8 bits?

This is about the length of a pretty complex proof, and the encoding of such a problem might be impossible even if $P \equiv NP$.

8.6 Notes

This appeared as why-believe-that-pnp-is-impossible in the blog. I would like to give a special thanks to Paul Beame, Jonathan Katz, and Michael Mitzenmacher for their kind help with this chapter. As always any errors are mine, and any good insights are theirs.

The discussion here was very lively and I cannot reproduce it. I would like again to thank all those who made comments. Thanks to Boaz Barak, Rich DeMillo, Steve Flammia, Timothy Gowers, James Martin, Torsten Palm, Jonathan Post, John Sidles, Aaron Sterling, and everyone else. Thanks.

Chapter 9
A Nightmare About SAT

Sigmund Freud was not a famous computer scientist. He is of course famous for being "the father of psychoanalysis," and pioneered the study of dream symbolism, especially nightmares. I have a recurring dream about P=NP that you might characterize as a "nightmare." I hope by telling you my nightmare, I will stop having it–perhaps my telling it will be therapeutic.

9.1 The Nightmare

There are several versions of my nightmare, but they all concern whether or not the complexity of NP can vary wildly. Suppose that we focus on the boolean circuit complexity of SAT. The same comments can be made about uniform complexity, but they are more straightforward for circuit complexity so we we will only consider circuit complexity.

Let $S(n)$ be the size of the optimal boolean circuit for SAT with an n bit description. Conventional wisdom says that $S(n)$ is super-polynomial, i.e. that $S(n)$ is larger than any polynomial n^k for large enough n. We talk about what happens when $S(n) = O(n^k)$ for some fixed k in another post.

The troubling possibility, to me, is that $S(n)$ could vary wildly as n tends to infinity. For all we know it could be the case that *both* the following are true:

- $S(n) < n^2$ for infinitely many n;
- $S(n) > 2^{\sqrt{n}}$ for infinitely many n.

This would be a terrible situation, since then sometimes SAT is "easy" and sometimes SAT is "hard." At first glance you might think that this is impossible. If SAT is easy infinitely often, then how could it be the case that the method that works on some n cannot be made to work on other n's. The nightmare is that this seems completely possible. There seems to be no way to argue that it cannot happen.

One can easily prove if $S(n)$ is small for some n, then certainly it must be small for all n nearby. More precisely, $|S(n) - S(n+1)|$ cannot be very big. This follows directly from the ability to pad a set of clauses. Thus, if $S(n)$ varies from easy to hard, the corresponding places must be far apart. But that is the only restriction that I can see on the behavior of the function $S(n)$.

9.2 Open Problems

An important open problem, in my opinion, is to resolve this question about SAT. Call the behavior of a problem *wild* if the corresponding complexity function varies greatly as n varies. Can we prove, even on some unproved assumptions, that $S(n)$ cannot vary wildly? This seems to be a very difficult problem.

Here is another problem that might be more attackable. Define $M(n)$ as the number of arithmetic operations that an optimal computation needs to perform $n \times n$ matrix multiplication. Can we prove that $M(n)$ cannot be a wild problem? The reason this seems possible is the potential to use the famous insight of Strassen: the recursive structure of matrix multiplication. I think this question might be resolvable. A concrete version would be to show that if $M(n) < n^\omega$ infinitely often for some constant ω, then $M(n) < n^\omega \log n$ for all n large enough. Clearly, there is nothing special about the function $n^\omega \log n$, one can replace it by others. Also the same question makes sense for other problems.

9.3 Notes

This appeared as a-nightmare-about-sat in the blog.

Chapter 10
Bait and Switch

Steven Rudich was the 2007 co-recipient of the famous Gödel Prize for his work on "natural proofs" [117]. In addition, Steven has done some beautiful work in the foundations of computational complexity. Lately, he has turned his considerable talents to education of students. He is a wonderful explainer of ideas and I wish I could be half as good a lecturer as he is.

Steven once explained to me why he thought that lower bounds are so hard to prove. His idea is really quite simple. Imagine that you have a friend who is interested in computing some boolean function $f(x_1, \ldots, x_n)$. Your job is to try to convince your friend that no matter how they proceed the computation of f will take many steps. Steven's point is that one way you might argue to your friend is: "Look—any computation that computes f must make slow progress toward f." Each computational step can only get you a little bit closer to the final goal, thus the computation will take many steps. This sounds like a plausible method.

Here is a more precise way to state this method. Let us define some measure on how close a boolean function g is to f. Then, we would show that each variable x_i is very far from f. The key would be next to show that if g and h are far from f, then $g \vee h$ cannot be much closer. (And the same for the other boolean operations of "negation" and "and" and "exclusive-or.") The fundamental idea is that as one computes the best one can do is to slowly move closer to f. The rub so far has been the difficulty in finding how to measure the distance between boolean functions.

This method makes a lot of sense to me. Any computation starts with the the lowly variables that must be far away from the goal of f. Then, as each step of the computation is made, the computation gets closer to the goal of f. This has a nice geometric flavor: we can think of the computation as taking a "walk" from variables toward the goal. Since f, the goal, is far away, this process takes many steps and we have a lower bound method. Sounds great. The trouble is that it cannot work.

10.1 Bait and Switch

Steven's brilliant insight is what I call a "bait and switch" trick. Bait and switch is my term–do not blame Steven. It wipes out most attempts that researchers have tried to use to prove lower bounds. I think that it is a very powerful tool for seeing that an approach to proving lower bounds is doomed to fail.

Recall that a bait and switch is a form of fraud. The idea is to lure customers into a store by advertising a product at a ridiculously low price, and then attempt to get them to buy a different product. The original product may not have even existed.

We can use the same method of fraud to "shoot down" lower bound attempts. Imagine again that we want to show that f is hard to compute, say that it takes at least n^2 steps to compute. We pick a random boolean function r that requires n^2 steps to compute. There are lots of these. Now we compute f as follows:

- First we compute $a = f \oplus r$;
- Then, we compute $b = r$;
- Finally, we compute $a \oplus b$.

Clearly, the final answer is f since $a \oplus b = (f \oplus r) \oplus r = f \oplus (r \oplus r) = f$.

I think of this as a bait and switch trick: you first compute $f \oplus r$ which has nothing to do with f. Also the same for r. Neither reveals anything about the function f. They are the bait. Then, at the very *last step* of the computation you put a and b together to form f. This is the switch step.

The reason this destroys lower bound methods is that computing a and b by themselves has nothing to do with f. Nothing at all. So intuitively how can any measure of progress make sense. How can we say that computing $a = f \oplus r$ is making any progress towards f? The function a is just some random boolean function of about the same complexity as f. So how can we measure slow progress towards f. Then, the switch of computing $a \oplus b$ magically reveals the "fraud." We have actually been computing f, but this is only apparent at the last step of the whole computation.

I find Steven's idea very compelling. I think that he has hit on the core of why lower bounds are so elusive. About two years ago I was on a program committee that received a paper claiming a major lower bound result. The paper was hard to read, and while we doubted the result we could not find any flaw.

Outside experts who had been reading the paper for a while, such as Steve Cook, had no idea if the paper was correct or not. The committee was in a bind: we could reject a potentially famous paper or we could accept an incorrect paper. I applied the "bait and switch" rule to the paper. The author had used a type of progress measure. He argued that as the computation proceeded at time t the algorithm must have tested at least some $d(t)$ of certain cases. Clearly, this would not pass the bait and switch trick. I felt confident in voting against the paper based on Rudich's insight. I did not see how the author's method could avoid the bait and switch trick. We rejected the paper. Since then the author has withdrawn his claims, so we were correct.

10.2 Open Problems

The obvious open problem is to try to build Steven's insight into a theorem. Can we prove that there is no measure approach to showing lower bounds? This seems possible to me. I cannot yet see how to make this into a formal theorem, however. Note, also these comments about bait and switch work for arithmetic computations too. We just replace $f \oplus r$ by $f - r$. All the same comments apply.

10.3 Notes

This appeared as bait-and-switch-why-lower-bounds-are-so-hard in the blog. See the next chapter for more on bait and switch.

Chapter 11
Who's Afraid of Natural Proofs?

Alexander Razborov is one of the world experts on circuit lower bounds. Perhaps the expert. In 1990 he won the Nevanlinna Prize for his work on lower bounds, more recently in 2007 he won the Gödel Prize (with Steven Rudich) for their beautiful work on "Natural Proofs" (NaP) [117].

His joint work on NaP is the subject of this chapter. Should we be worried about the so-called "Natural proof obstacle" to resolving P=NP? Or is this obstacle not an obstacle? I do not know, but would like to discuss this issue with you.

Do "impossibility proofs" matter? I mean proofs that show that "you cannot solve problem X by method Y"? Sometimes they do: the Gödel Incompleteness Theorem is a perfect example, it essentially destroyed any attempt to complete David Hilbert's program and prove arithmetic consistent. Yet, Gerhard Gentzen carried on and proved that first order arithmetic is consistent, although he had to use transfinite induction up to ε_0. This is the limit ordinal of the sequence,

$$\omega, \omega^\omega, \omega^{\omega^\omega}, \dots$$

The great logician Alfred Tarski on hearing of Gentzen's Theorem was asked if he was now more confident of arithmetic's consistency–he answered "by an epsilon." (Gentzen's work was based on a logic that he called "natural deduction"–curious coincidence?)

Another example is the great work of Paul Cohen. I met Cohen at a Royal Society workshop on the notion of "mathematical proof." It focused on the growing number of computer aided proofs, and what the math community should think about proofs that are not checkable by humans. There is now a growing list of such proofs; first there was the four color theorem, then Kepler's problem, then again the four color theorem, then some results in geometry. The list is growing.

Cohen is, of course, world famous for creating the notion of forcing, and for proving that the Axiom of Choice and the Continuum Hypothesis are independent from ZF set theory [35]. After he proved this great result he was asked if he had been worried about his approach, since it came very close to violating a known impossibility theorem that had been proved years before. He—it is claimed—said, "what result?" I find this story, true or not, interesting. We must not let impossibility theorems scare us into not thinking about lower bounds. We can let them guide us: we can let them show us what we cannot do, and what we should do. But we must move forward.

11.1 The Obstacle

The obstacle to lower bounds is this. Suppose that you want to prove that a boolean function f has high complexity–in some sense. Then, you might think that you could use special properties of this particular boolean function. Makes sense. Perhaps the function is the sign of the determinant of a square matrix. The determinant has many special properties that perhaps you can use in your lower bound proof: for example the determinant of a invertible matrix is non-zero, or that certain determinants count the number of spanning trees of a graph, and so on.

The obstacle results show, roughly, that a large class of approaches to circuit lower bounds must prove more. Your lower bound for f must also prove that many other *unrelated* functions also have large complexity. Thus, you cannot use any special properties of your function. None. This seems like a major obstacle. Imagine that I ask you to prove that a certain number is prime, but your proof must work for many other numbers. That sounds nuts. When we prove that a number is prime we prove that and nothing else.

I think of the obstacle as a kind of *drag-along principle*: if you prove f is complex, then you prove much more.

11.2 The Obstacle One: Bait and Switch

There are two concrete theorems that show different versions of the drag-along principle. I like to call the first one "Bait and Switch": it is close to a question that I posed in Chp. 10. However, it is a bit different from what I had in mine at the time: I was looking for a stronger result.

Let B be the set $\{0,1\}$. Then, we plan to study the circuit complexity of functions $f : B^n \to B$. We use $\mathscr{B}_n$ to denote the space of all boolean functions with n inputs, i.e. $f : B^n \to B$. We also use $\mathscr{M}_n$ to denote the space of all n-input monotone functions.

If f is n-input boolean function, i.e. $f \in \mathscr{B}_n$, then we use $C(f)$ to denote the size of the smallest circuit for f: in our circuits all the negations are at the inputs and we only allow $\{\vee, \wedge\}$ as operations. We use $C_m(f)$ to denote the monotone complexity of f for $f \in \mathscr{M}_n$.

I follow the treatment of Sanjeev Arora and Boaz Barak [8]. The key idea is to look at *complexity measures* on boolean functions. A measure $\mu(f)$ is a *complexity measure* if it assigns positive values to each boolean function. It is normalized so that it assigns 1 to variables and their negations. Also it must satisfy the following two simple rules:

$$\mu(f \vee g) \leq \mu(f) + \mu(g)$$

and

$$\mu(f \wedge g) \leq \mu(f) + \mu(g).$$

As they point out, a lower bound on $\mu(f)$ yields a lower bound on the size of the formula complexity of f: this follows by a simple induction.

The Bait and Switch Lemma (BSL) is the following:

Lemma 11.1. *Suppose μ is a complexity measure and there exists a function $f \in \mathscr{B}_n$ such that $\mu(f) > s$. Then, for at least $1/4$ of all g in $\mathscr{B}_n$, $\mu(g) > s/4$.*

Proof. Let g be any function in $\mathscr{B}_n$. Define $f = h \oplus g$ where $h = f \oplus g$. Then, $\mu(f) \leq \mu(g) + \mu(\bar{g}) + \mu(h) + \mu(\bar{h})$: this follows from the definition of exclusive "or", since

$$f = (f \oplus g) \oplus g = h \oplus g = h \wedge g \vee \bar{h} \wedge \bar{g}.$$

By way of contradiction assume that $\{g : \mu(g) < s/4\}$ contains more than $3/4$ of all $\mathscr{B}_n$. If we pick the above function g randomly, then $\bar{g}, h, \bar{h}$ are also random functions (though not independent). Using the trivial union bound we have

$$Pr[\text{All of } h, \bar{h}, g, \bar{g} \text{ have } \mu < s/4] > 0.$$

This implies by the definition of complexity measure that $\mu(f) < s/4$, which is a contradiction.

The reason this is an interesting theorem is that it shows that any lower bound based on a complexity measure would have to prove more. It would have to prove that not only f but many other functions are hard. Thus, it would not be able to use the special structure of the boolean function f.

Well I must point out that about twenty years ago, a very famous complexity theorist circulated a proof using exactly this method to show that SAT had super-polynomial boolean complexity. The proof of course was wrong, but it took "us" a while to find the bug. No it was not me, and I will not say who.

11.3 Obstacle Two: Natural Proofs

The second drag-along principle is NaP. I will not explain the details of the notion of NaP's. The main idea is suppose that we wish to prove that a boolean function $f \in \mathscr{B}_n$ has large complexity. A "natural" idea is to find a predicate Φ on boolean functions in $\mathscr{B}_n$ so that (i) $\Phi(f)$ is true and (ii) $\Phi(g)$ is true for a large fraction of $\mathscr{B}_n$.

Then, it should follow that $C(f)$ is large. This is reasonable, since the conditions roughly say that f has some key property in common with random boolean functions. Standard counting methods prove that most boolean functions are hard.

Making all this work is the brilliance of the NaP paper. Note, they need a stronger condition on the predicate $\Phi(g)$ and other assumptions. See their paper for the details [117].

They show most (all?) of the standard lower bound proofs follow the above structure. That is the reason that NaP seems to such a powerful idea. You are forced to drag-along too much.

11.4 Monotone Circuits

I wonder if the following can avoid the drag-along obstacle: use monotone complexity. The idea is simple, in order to get a lower bound on a boolean function f, for the "right" g,

prove a monotone lower bound on g

then

get a general lower bound on f.

To carry out this program we will study an old idea (of mine) that defines a *transformation* on boolean functions that has several interesting properties. If f is a boolean function, then $Ł(f)$ is a monotone boolean function where $Ł$ is our transformation. Moreover, its monotone circuit complexity is almost the same as f's non-monotone circuit complexity.

You can *skip* the rest of this section if you believe what I just said and do not need to see any details. Or you can read on. Or both. Your choice.

Now let us turn to defining our transformation. We need two simple auxiliary functions α and β. Let $\alpha(x_1,\dots,x_n,y_1,\dots,y_n)$ equal

$$\bigwedge_i (x_i \vee y_i)$$

and let $\beta(x_1,\dots,x_n,y_1,\dots,y_n)$ equal

$$\bigvee_i (x_i \wedge y_i).$$

Let $f : B^n \to B$. Then, $Ł(f)$ is equal to

$$\alpha(x_1,\dots,x_n,y_1,\dots,y_n) \wedge f(x_1,\dots,x_n) \vee \beta(x_1,\dots,x_n,y_1,\dots,y_n).$$

Note, that $Ł(f)$ is a function of $2n$ variables, i.e. $Ł(f) : B^{2n} \to B$. The main properties of our transformation are contained in the next theorems.

Theorem 11.1. *For any boolean function f, $Ł(f)$ is a monotone function.*

Proof. Suppose that f is an n-input boolean function. Let $Ł(f)$ be equal to $g(x,y)$. If g is not monotone, there must be a,b so that $g(a,b) = 1$ and changing some input from 0 to 1 makes $g = 0$. Clearly, $\beta(a,b) = 0$; otherwise, after the change β would still be 1. Since $g(a,b) = 1$ it must be the case that $\alpha(a,b) = 1$. But then after the change β must be equal to 1. This is a contradiction.

Theorem 11.2. *For any boolean function f,*

$$Ł(f)(x_1,\ldots,x_n,\neg x_1,\ldots,\neg x_n) = f(x_1,\ldots,x_n).$$

Proof. Suppose that $f(x_1,\ldots,x_n)$ is a boolean function. Let y be equal to $\neg x_1,\ldots,\neg x_n$. Then, by definition $\alpha(x,y) = 1$ and $\beta(x,y) = 0$. Thus, $Ł(f)$ is equal to $f(x)$.

Theorem 11.3. *For any boolean function f,*

$$C(f) \leq C_m(Ł(f)) \leq C(f) + 4n.$$

Proof. Suppose that f is an n-input boolean function. $C(f) \leq C_m(Ł(f))$ follows from Theorem 11.2. Now suppose that f has a circuit of length l: let the circuit for f be $f_1,\ldots,f_l$. We will construct a monotone circuit for $Ł(f)$. Use $x_1,\ldots,x_n$ and $y_1,\ldots,y_n$ as the inputs. To start let $g_1,\ldots,g_l$ be the circuit where g_i is constructed as follows:

1. if f_i is equal to the input x_j, then g_i is the same input x_j;
2. if f_i is equal to the negated input $\neg x_j$, then g_i is the input y_j;
3. if $f_i = f_j \vee f_k$, then $g_i = g_j \vee g_k$;
4. if $f_i = f_j \wedge f_k$, then $g_i = g_j \wedge g_k$.

Next compute $g_m = \alpha \wedge g_l \vee \beta$:

$$g_1,\ldots,g_l,\ldots,g_m.$$

This takes at most $4n$ extra steps. So the total computation is now $l + 4n$ long.

The key is the following claim: $g_m = Ł(f)$. Suppose that g_m is different from $Ł(f)$ for some values of the inputs x and y. Then, clearly, $\beta(x,y) = 0$; otherwise, they would agree. Also $\alpha(x,y)$ must equal 1; again, if not, the two values could not disagree. We now claim that for each k, $x_k = \neg y_k$. Suppose that this was false. Then, let $x_k = y_k$ for some k. Clearly, the common value cannot be 1 since $\beta = 0$. Also the common value cannot be 0 since $\alpha = 1$. This proves that for each k, $x_k = \neg y_k$.

Finally, a simple induction proves that $g_i = f_i$ for $i \leq l$. This is clear since the only difference in the two computations is that y_k takes the place of $\neg x_k$. This shows that $g_l = f$ and that $g_m = \alpha \wedge f \vee \beta$.

All the above can be generalized to handle formula complexity instead of circuit complexity. There is essentially no difference in the proofs.

11.5 Open Problems

My question that I cannot resolve is does the following violate either the Bait and
Switch Lemma (BSL) or the NaP result. Suppose that f is a boolean function that
you wish to show has high complexity. Prove that the monotone complexity of $L(f)$
is large. Then, use our theorem to conclude that the general complexity of f is large.

Bait and Switch Lemma: This does not seem to obviously apply. The problem
is that we get a lower bound on $Ł(f)$ which is a monotone function. There does
not appear to be a BSL for monotone functions. The problem is how can we "add"
something to a monotone function and then remove it later on?

Natural Proofs: This also does not seem to directly apply. The problem now is
that $Ł(f)$ is highly not random. It is a monotone function, but even more: it is almost
always 1. As a matter of fact, it will be unequal to 1 with probability at most 2^{-n}.
How can a function that is essentially a constant look like a random one?

I think these questions may be easy to resolve, and that the two obstacles are
indeed still obstacles. But I am confused.

11.6 Notes

This appeared as whos-afraid-of-natural-proofs in the blog. Timothy Chow discussed whether NaP could be used to show the difficulty of proving even non-linear
circuit lower bounds.

Chapter 12
An Approach To P=NP

Dexter Kozen is a famous theorist, who among many other wonderful things co-invented the notion of alternating Turing Machines. He also is the author of a terrific book [84]—his other books are pretty neat too—but this one is my favorite. Dexter also proved a *controversial theorem* decades ago about the power of diagonal methods in the separation of complexity classes. More on that later.

I want to talk about a possible approach to the separation of powerful complexity classes. The approach is based on a diagonal argument, as predicted by Dexter's paper.

When alternating Turing Machines were discovered, the idea was somehow in the "air." Two papers at the same FOCS presented essentially the same results; these papers were later merged into one famous paper by Ashok Chandra, Dexter Kozen, and Larry Stockmeyer [30].

I was at Yale, in the computer science department, when Dexter co-invented alternation. It's a long time ago, but I was working with a graduate student named David Cotton. He was, one of the few students who I worked with over the years that dropped out of the Ph.D program—I somehow failed him. He had some family issues that probably made it hard for him to continue, but I always felt bad that he did not get his degree. I have lost touch with him, and hope he is doing well.

I mention David because just before FOCS that year, David was beginning to work on alternation. We did not call it that, of course, but as I said the idea was in the "air." He was working on relating alternating time and space, when we heard about the FOCS results. We did not have the results, but I think that David was getting close, since he was a very strong problem solver.

An example of his ability was evident one day at a problem seminar that we ran at the time at Yale. The seminar meet once a week, where we discussed open problems and people presented ideas or even solutions to the problems.

One week someone, I forget who, presented a neat problem about finding a linear time algorithm for a certain problem—let's just call it X. The plan was to meet the next week and see if anyone had solved the problem. David did not attend the meeting when we discussed X. Since this was pre-email and Internet, I do not think he even heard about the problem.

That week I spent a huge amount of time trying to get a linear time algorithm for the problem. Partly, X was a cool problem, and mostly it was the competitive spirit. I wanted to solve the problem, before anyone else did. I eventually found a complex recursive algorithm that seemed to be linear, but the analysis of the running time depended on solving a tough recurrence—one that was too hard for me to solve. So I when down the hall to see Stan Eisenstat, who is a numerical analyst, and can solve just about any recurrence. He convinced me quickly that my algorithm ran in $n \log n$ time, and I stopped working on the problem—I had no other ideas.

At the next meeting of the seminar, Stan started to explain a linear time algorithm that he had "found." I quickly saw that Stan was presenting "my" algorithm, but he had changed his mind and now could prove that it ran in linear time. However, the analysis was really complex, soon the blackboard was covered in mathematics: a lemma here and an equation there. It was a mess.

Then, David walked in; he always arrived late for the seminar. He saw Stan at the board and asked someone what was he doing. They handed David a sheet that described the problem X that Stan was trying to solve. David said to no-one in particular, "but that is trivial." Stan ignored him. Then, David said it louder, and finally Stan turned to David, and as Stan slowly put his chalk down, in the chalkboard tray, he said with a smile on his face:

so why don't you come up and show us *your* solution.

David stood up, walked up to the board, picked up the chalk, and proceeded to present an algorithm that ran in time $2n - 1$. The algorithm, once you saw it, was obviously correct—we all had missed it. Stan was speechless, as we all were. The meeting broke up.

Too bad David moved on to other things. In any event, let's turn to see how we might solve P=NP. Perhaps if David was around still, perhaps he could help us make the following ideas work.

12.1 Upper Bounds That Imply Lower Bounds

I would love to see us start to prove some serious separation results. I know I have argued for possible collapses, but I do believe that some of our complexity classes may be different. So I decided that it was time to write about an attack on separation theorems.

Here goes. The main issue that I see is that lower bounds are very difficult. You then say, "how can we then prove separation results without proving lower bounds?" The answer is based on using simple diagonalization along with some combinatorial cleverness. The hope is that this method will be able to avoid the known pitfalls—such as oracle results.

The method is based not on finding lower bounds, but on proving upper bounds. A simple analogy is: suppose I want to show you that $X < Y$. I can do two things: I can prove that X is small, and conclude that X must be less than Y; or I could prove that Y is large and conclude that Y is larger than X. They are really different approaches.

Here is an old example I discuss in chp. 6. If all sets in P have quadratic size circuit complexity, then a neat result of Ravi Kannan shows that P$\neq$NP. More generally,

Theorem 12.1. *Suppose there is a constant $c > 0$ so that for any $S \subseteq \{0,1\}^*$ computable in polynomial time, for each n the circuit complexity of $S \cap \{0,1\}^n$ is $O(n^c)$. Then, P does not equal NP.*

Proof. By way of contradiction assume that P=NP. Ravi's theorem proves that for any d there is a set in second level of the polynomial hierarchy that requires a circuit of size n^d for all n. But, since we have assumed that P=NP, the hierarchy collapses: thus, there is a set S_d that is in P, yet S_d requires circuit complexity of n^d. This is a contradiction, since we are free to choose d so that $d > c$.

12.2 Conditional Upper bounds and Separation

The basic insight is that we are good at creating algorithms, and terrible at proving that algorithms *do not exist*. My approach to separation is to show that the following type of implication can be proved:

$$\mathscr{U} \implies \mathsf{L} \neq \#\mathsf{P}.$$

Where the statement $\mathscr{U}$ is an assertion that a certain upper bound is true, conditioned on the assumption that $\mathsf{L} = \#\mathsf{P}$.

The conditional assumption is "free," since we can assume it and reach a contradiction. Thus, the upper bound $\mathscr{U}$ can use any consequence of the collapse $\mathsf{L} = \#\mathsf{P}$. In particular the upper bound can use guessing and counting as part of the algorithm. Note, this follows since the collapse makes all of these fall into the class L.

I have several upper bounds in mind. I will present the simplest one that is based on the determinant. There are technically weaker statements that I believe can be used. Of course the ultimate goal is to try and get an upper bound assumption that is *provable*. Such a bound, would lead us to an unconditional separation of the classes L and $\#\mathsf{P}$.

Note, before we go further that there is one interesting difference in this approach from trying to find a circuit *lower bound*. We currently have no idea how to prove strong circuit lower bounds in the general model. Even worse there is evidence that they may be hard to prove—for example, the work on Natural Proofs. On the other hand, there is no reason, that I know, to believe that clever upper bounds, especially conditional ones, exist for reasonable problems. We need only apply our algorithmic creativity to show that the right $\mathscr{U}$ is true and then we are done.

We are trying to prove implications of the form:

$$\mathscr{U} | \neg S \implies S.$$

Here $\mathcal{U}|T$ means the statement $\mathcal{U} \wedge T$: I like the notation since it emphasizes the fact that T is being conditionally assumed.

12.3 What To Try To Prove?

I think that P=NP is possible, but if I had to try to prove that they are unequal I would proceed as follows. I would try to first prove something much "easier," well at least not as hard as the full problem. What I would do is try to prove that

$$L \subsetneq \#P.$$

It is entirely possible that #P could equal PSPACE so that the above could certainly be true.

This reminds me of a great talk given by Juris Hartmanis at Yale back in the 1970's. During the talk, he put up a slide that showed,

$$L \subseteq P \subseteq NP \subseteq \#P \subseteq PSPACE.$$

He then pointed out that at least one of the subset relationships must be strict, but we have no idea which one. The reason they must be strict is that L is easily shown by diagonalization to be weaker that PSPACE: thus, they all cannot be equal. But which subsets are strict?

12.4 Describing Circuits

We need a notion of non-uniformity. Suppose that A is a circuit and $x = x_1, \ldots, x_n$ is a string. Say that A *describes* x if for all $i = 1, \ldots, n$,

$$A(i) = x_i.$$

Let us use $A \to x$ to denote this.

Now suppose that S is a set. We use $S \in TC^0//f(n)$ to mean that that there is a depth d and a polynomial n^c, so that for all n there exists a general boolean circuit A of size $f(n)$ with the following property: There is a threshold circuit T of depth d and size at most n^c such that

1. The circuit A describes T, that is $A \to$ Encode(T) where Encode(T) is some natural encoding of the threshold circuit T;
2. The threshold circuit T computes S for all inputs of length n.

By $\det_n$ let us mean the boolean function of deciding the determinant of an n by n matrix, where the entries are from the finite field of two elements.

12.5 How To Try To Prove It?

Here is a sample of the type of result we are working on:

Theorem 12.2. *Suppose that computing* $\det_n$ *is in* $TC^0//n^{\varepsilon}$ *for any* $\varepsilon > 0$. *Then,* $L \subsetneq \#P$ *is true.*

I am working on the proof of this with Ken Regan. The rough outline is this. Assume by way of contradiction that L=#P—call this the "collapse." Then, suppose that M is a Turing machine that runs in time say n^{100}. The actual running time will depend on the constant in the collapse of #P to L—but, for the outline we will ignore this. The goal is to simulate M in time n^{99}, which will violate the classic time hierarchy theorem.

How do we do this? Well we have a number of powerful tools that we can use:

- We have the collapse, which allows us to "count" the size of any easy-to-describe set. This is very powerful.
- We also have that the determinant can be computed by *some* circuit that has constant depth and polynomial size. Note, we do not know which circuit it is, but, we do know that there is some circuit.
- We can "find" the circuit by using the collapse again, since the collapse allows us also to do guessing efficiently. Of course, we must be able to check that our guess is correct.
- Finally, we can reduce the simulation of M to determining whether or not a graph is connected. This can be done by calculating a determinant by the famous Matrix Tree Theorem. Note, we are using the following fact: a graph is connected if and only if it has a spanning tree, which in turn is true if and only only if a certain matrix has a non-zero determinant.

12.6 Open Problems

Can we make this work? We actually think we can weaken the hypothesis on the determinant quite a bit. Roughly we only need that an approximate determinant type function can be computed.

12.7 Notes

This appeared as an-approach-to-the-pnp-question in the blog. Ryan Williams had a nice observation about a possible way to solve some of the questions raised.

Chapter 13
Is **SAT** Easy?

Ed Clarke, a Turing Award winner, was at a recent NSF workshop on design automation and theory. Ed has, for approximately three decades, worked successfully on advancing the state of the art of verification, especially that of hardware.

Let's talk about a discussion that came up at the workshop. Ed and others stated that in "practice" all the instances of **SAT** that arise naturally are easily handled by their **SAT** solvers. The questions are simple: why is this true? Does P=NP? What is going on here?

I have known Ed since he presented his first paper [32] at the 1977 POPL, a conference that I used to attend regularly. He proved a wonderful theorem about an approach to program correctness that was created by Steve Cook. I worked on extending Steve's and Ed's work, my work appeared later that year [94].

Cook had a clever–not surprising–idea on how to circumvent the inherent undecidability of determining anything interesting about programs [40]. I will try to motivate his idea by the following program:

```
Search all natural numbers x,y,z ordered by
the value of x+y+z,
  if f(x,y,z)=0, then print x,y,z and stop.
```

Here f is some polynomial. Steve's insight was the reason this program is hard to understand has nothing to do with the complexity of the program. Rather, it is all due to the domain complexity of the natural numbers.

Cook's brilliant idea was to figure out a way to "factor" out the domain complexity. He did this by adding an oracle that would answer any question about the domain. In the above program, this ability means that telling if the program halts is easy: just ask the oracle "is there an x,y,z so that $f(x,y,z) = 0$?"

Sounds like a trivial idea. But, wait. If you have access to an oracle for the natural numbers, then anything is easy to determine about any program. This would lead to a trivial theory. Cook had to rule out the possibility that the oracle would be asked "unnatural questions." By this I mean that you might be able to use the fact the arithmetic is very powerful and encode the analysis of your program into an arithmetic formula. Cook did not want to allow this.

The way he did this is cool: the analysis of the program has access to the domain oracle, but the analysis must be correct for *all* oracles. Thus, if you try to encode information in some "funny way", then for some domain(s) your analysis is wrong. This was the clever insight of Cook.

Steve showed that certain kind types of programs could be analyzed with such an oracle and any domain, while Ed proved that certain more powerful type programs could not. I proved a necessary and sufficient condition on the type of programs for which Steve's method works. My proof is quite complex–I try not to say that about my own work–but it really is tricky. Look at the papers if you are interested.

13.1 Thanks Paul

Before we turn to SAT solvers, I want to thank Paul Beame who has helped tremendously with this. I thank him profusely for his time and effort. Think of this a guest post, but all the errors and mistakes are mine. Anything deep is due to Paul.

13.2 SAT is hard, SAT is easy, SAT is hard,...

There is something fundamentally strange going on. The theory community is *sure* that P≠NP, and so SAT is hard. The solver community is *sure* that they can solve most, if not all, of the problems that they get from real applications, and so SAT is easy. What is going on here?

In Chp. 3, I raised some of the many ways that the P=NP question could play out. There are plenty of ways that both the theorists and the solvers could be right. As you know I like to question conventional wisdom, and in particular I am not convinced that P=NP is impossible. But, I also question the conventional wisdom of the solvers that *practical* SAT problems are all easy. So I feel a bit like we are living in Wonderland with Alice: (with apologies to Lutwidge Dodgson aka Lewis Carroll.)

> ""There is no use trying; one can't believe impossible things, like P=NP." Alice says.
> "I dare say you haven't had much practice. When I was your age, I always did it for half an hour a day. Why, sometimes I've believed as many as six impossible things before breakfast, like L=P, P=NP, NP=BQP, and ..." The Queen replies.

13.3 SAT Solvers

In order to start to understand the situation let me first discuss the state of the art in SAT solvers. I have used many comments from Paul, and also have relied on the pretty survey paper [57] of Carla Gomes, Henry Kautz, Ashish Sabharwal, and Bart Selman. This is definitely not my area of expertise, but I hope I can get the main points right. As always all the errors are mine, and all the correct stuff is due to Paul or my inability to transfer the facts correctly from the survey. Please let me know if I made any mistakes.

My understanding is SAT solvers divide roughly into two classes: complete and incomplete. Complete means that the algorithm always finds an assignment if the problem is satisfiable; incomplete means the algorithm may not. The complete ones all use the procedure of Martin Davis, Hilary Putnam, George Logemann, and Don Loveland (DPLL) as a basis, but usually add sophisticated methods to make the algorithm go faster. The incomplete methods tend to be different variations on random walk methods. I've known the D and second L well in DPLL: I knownd Davis for a long time and Loveland taught me logic while I was a graduate student at CMU.

The DPLL method is based on a backtracking method. It assigns literal values and makes recursive calls to the simplified set of clauses. If it gets stuck, then it backtracks and continues searching. This method was initially called the DP method, until Logemann and Loveland added two additional insights: take advantage of unit clauses and "pure" literals. If a literal x occurs, for example, always as x and never as $\bar{x}$, then all clauses with x can be removed. Such a literal is called pure. Paul points out that the history and relationship from DP to DPLL is more complex, but I will leave that out.

Paul says that there are three keys to fast solvers today.

1. *Branching heuristics*, which take into account where the action is based on the learned clauses. One method is called VSADS and uses a weighted majority algorithm.
2. *Clause learning*, which is a way of adding new clauses to the formula based on failed branching. This can potentially help avoid large segments of the search.
3. *Watched literals*, which allow the modified formula to be kept implicit. This makes the propagation of the updates fast and tends to keep everything in the cache.

Like all backtracking methods, the selection (1) of which literal to set next is extremely important. The same is true, for many other search algorithms. For instance, linear programming with a good pivot rule seems to take polynomial time in practice. Another example, is searching game trees with $\alpha - \beta$ search. If the "best" move is always used first, then the search tree is dramatically decreased in size.

There are many additional ideas that are covered in the survey. One key idea is to make random decisions when to restart the backtracking. The method is restarted even if it is not stuck. The method usually keeps the clauses already set to guide the next restart (2) of the backtrack algorithm. Finally, note that idea (3) is an implementation one, which is of great importance on modern computers. No doubt similar methods will soon be needed due to the rise of many-core processors.

The second class of SAT solvers is based on random walk technology. These algorithms start at a random assignment and then use various local moves to change the assignment, hopefully toward a correct one. Some algorithms make local improvements, some allow plateau type moves, others use more complex types of random walks. There is evidence that these methods can work well on random SAT problems as well as some of those from real applications.

There also are methods that are random, but are fundamentally different. A relatively new one is called *Survey Propagation* (SP), which was discovered in 2002 by Marc Mézard, Giorgio Parisi, and Riccardo Zecchina. The solvers based on SP are like belief propagation methods that are used for decoding error-correcting codes [22]. In each iteration about 10% of the literals are set. When the clause density gets too low, SP switches to a random local search algorithm. Paul points out that their article seems to not say anything about this part of SP.

13.4 Why are Real SAT Problems Easy?

The solvers know well that any DPLL type method cannot be fast on all SAT problems, based on known lower bounds for resolution. However, that still does not explain why these methods seem to work so well in practice. Even the contests that are held regularly on solving SAT divide the problems given to the solvers into three classes:

- randomly generated sets of clauses;
- hand crafted sets of clauses;
- examples from "real" applications.

Paul points out that he would take Ed Clarke's contention that "SAT is easy in practice" with a grain of salt. First, before SAT the verification community was trying to solve PSPACE-hard problems, so SAT is a great advance since is *only* NP-complete.

Second, the SAT solvers work except when they fail. This reminds me of the magician and parapsychology investigator James Randi's quip about ESP: "it works except when it fails."

The real difficulty is that no one understands what causes a SAT solver to fail. Tools that have this type of unpredictable behavior tend to create problems for users. Many studies of human behavior show that it's the variance that makes tools hard to use.

Paul goes on to say that one of the research directions is the discovery of why SAT solvers fail. One problem is symmetry in the problem. Problems with large amounts of symmetry tend to mess up backtracking algorithms. This is why the Pigeonhole Principle is a "worst case" for many SAT methods.

Even if solvers cannot do as well as Clarke claims, I still think that one of the most interesting open questions for a theorist is why do the algorithms do so well on many real problems? There seem to be a number of possible reasons why this behavior is true. I would love to hear about other potential reasons.

- The real problems may have sufficient *entropy* to behave like actual random SAT problems. Michael Mitzenmacher suggested this idea to me. For example the SP method can handle over one million clauses of randomly generated sparse clauses, according to Selman.

- The real problems may have small *treewidth*. More precisely the graph that describes the interaction of the literals has small tree width. Suffice it for now that many search problems with treewidth w and size n can be solved in time order $n^{O(w)}$.

- The real problems may have some *modular structure*. Clearly, if a problem comes from modeling a hardware circuit, the SAT clauses may have special structure. They may not have small tree width, but they should have some additional structure. What this structure is and how it interacts with the SAT solvers is unclear. Luca Trevisan suggested this idea to me. One problem that I pointed out is that many crypto-systems, such as DES/AES, have a very structured hardware circuit. Yet the folklore, and published papers, suggest that SAT solvers do not break these systems. So the notion of modular SAT problems is perhaps yet to be found.

- The real problems may have been *selected* to be easy to solve. A hardware designer at Intel, for example, may iteratively refine her design so that the SAT solver can succeed in checking that it satisfies its specification. This is a kind of "natural selection": perhaps designs that cannot be checked by the solvers are modified. The closest thing that theorists study that has a bit of this flavor is famous *smoothed analysis* technology of Dan Spielman and Shang-Hua Teng [124]. Maybe for every practical design there is one "nearby" that can be checked efficiently. Suresh Venkatasubramanian suggested this idea to me. I think it raises some nice questions of how to model smoothed analysis in a discrete domain.

- The real problems may have … More ideas needed. Please feel free to suggest some other ideas.

13.5 Open Problems

I love quoting Alan Perlis about the power of exponential algorithms-see Chp.20. I will start soon to quote Paul Beame who has made some great comments on this issue. Somehow we teach our theory students that exponential algorithms are "bad." This seems quite narrow and theorists should look at the design and analysis of more clever algorithms that run in exponential time.

I also think another issue raised here is the reliance we place on the worst-case model of complexity. As I discuss in chp. 28 there is something fundamentally too narrow about worst case complexity. The fact that SAT solvers seem to do so well on real problems suggests to me a problem with the worst-case model. How can we close this gap between theory and practice?

Perhaps the main open problem is to get theorists and solvers more engaged so that we can begin to build up an understanding of what makes a SAT problem easy, and what makes it hard. And what are the best algorithms for "real" SAT problems.

13.6 Notes

This appeared as sat-solvers-is-sat-hard-or-easy in the blog. There was a lively discussion on whether SAT is easy or not. One of the common themes was perhaps
SAT problems that arise in practice are precisely those that can be solved. I thank all
those who made comments, and especially Paul Beame, John Fries, Robert Harper,
Rafee Kamouna, Ken Regan, John Sidles, and Sundar Srinivasan.

Chapter 14
SAT is Not Too Easy

Lance Fortnow is one of the top complexity theorists in the world, and has proved fundamental results in many aspects of complexity theory. One example, is his famous work on interactive proofs and the complexity of the permanent with László Babai, which paved the way for the later result of Adi Shamir that IP = PSPACE [12, 121]. More recently, he did a beautiful extension of a result of Boaz Barak on the BPP hierarchy problem. He is of course famous for the longest running and best blog on complexity theory, and while doing all this he found time to be the founding editor-in-chief of the new ACM Transactions on Computation Theory.

Lance has also has played a unique role, I hope he does not mind me saying this, in shaping modern computational complexity. He made a "mistake" in a paper with Mike Sipser that opened a new chapter in complexity theory: this was the discovery that amplification by parallel repetition is not trivial [50].

In science, especially mathematics, we all try to do correct work, but sometimes the right error can be very important. I would argue that Lance and Mike, by finding that it was not obvious that parallel repetition amplifies, made one of the great discoveries of complexity theory.

An "error" that leads to a great discovery has happened many times before in the history of science. One of the best examples, I think, might be the work of Henri Poincaré on the 3-body problem. Poincaré wrote a prize winning paper on solving the 3-body problem, then discovered that he had made a mistake, and finally discovered what we now call *chaos* [127]. A pretty good "mistake."

Lance has a great sense of humor and is a terrific speaker. Not too long ago he gave a talk at a theory conference without using any visual aids: not chalk, not marking pens, not overheads, not powerpoint, nothing. The talk went well, which says something about his ability to make complex ideas understandable.

I have the pleasure to have a joint paper with Lance, and I will talk about that work: various lower bounds on the complexity of SAT. Given that so many believe strongly that SAT is not contained in polynomial time, it seems sad that the best we can prove is miles from that. It is like dreaming that one day you will walk on the moon, but today the best you can do is stand on a chair and look up at it at night. We are so far away.

Perhaps this is a hopeful analogy, since we did eventually walk on the moon. It took many terrific people, some great ideas, a huge effort, and perhaps some luck, but eventually Neil Armstrong and Buzz Aldrin did walk on the moon. Perhaps one day we will show that SAT cannot be computed in polynomial time. Perhaps.

14.1 The Results

SAT is of course the problem of determining a truth assignment for a set of clauses, and is the original NP-complete problem. Yet what we know about the complexity of SAT is pretty slim. The conventional wisdom, of course, is that SAT cannot be recognized by any deterministic polynomial time Turing machine. What we believe is vastly different from what we can prove. The best results look like this:

Theorem 14.1. *No deterministic Turing machine accepts* SAT *in time* n^c *and space* $o(n)$ *where* $c \leq \lambda$.

Here λ is some fixed constant that is unfortunately small: I call it a "λ", since the first result was due to Lance.

Initially, it was shown by Lance that $\lambda > 1$. Then, it was proved by by Anastasios Viglas and myself that $\lambda \geq \sqrt{2} - \varepsilon$. Note, we made a mistake, but ours was "silly": for a short time we thought that we could prove $\lambda \geq 2$. Next Lance and Dieter Van Melkebeek proved that $\lambda = \phi - \varepsilon$ where ϕ is the golden ratio, i.e. $\phi \approx 1.618$. Ryan Williams then improved this to $\lambda \geq 1.7327$ which is just above $\sqrt{3}$ [134]. Even more recently it has moved up to 1.803 [135].

14.2 The General Structure

In his beautiful paper, Williams gives a general framework for all the known proofs. If you are interested in working on lower bounds for SAT, you should look at his paper–and ours too. I want to give you the 50,000 foot view of how such lower bounds are proved.

1. Assume by way of contradiction some complexity assumption that you wish to show is false;
2. Use this assumption to show that an algorithm exists that *speeds up* some computation;
3. Finally, show that this speedup is too fast: it violates some known diagonalization theorem.

Williams gives a much lower level description and further formalizes the methods to the point that he can even prove results on the limits of these methods, this is a unique aspect of his work. But for now I hope that this summary helps make some sense of these results:

assume something $\rightarrow$ get fast algorithm $\rightarrow$ show too fast .

What I find nice about this type of proof is that you get lower bounds by finding clever new algorithms. The algorithms typically must use some assumed ability, but they are nevertheless algorithms. One thing that Don Knuth has said is that computer science is *all* about algorithms. You might find that a bit extreme, but the key point is that this method allows you, actually forces you, to think in terms of new algorithms. You do not need to know some exotic mathematics, but you must be able to program Turing machines.

The other advantage of this method is that lower bounds you get are by definition general: the method shows that *no* Turing machine of a given type can do something. There are usually time and space restrictions on the Turing machines, but no other restrictions. They can do whatever they want within those constraints. Thus, these lower bounds are much stronger than monotone, or bounds of constant depth, or any other bounds that restrict computations in some way.

14.3 The Main Trick

In my view, there is one basic trick that we have used over and over in complexity theory. The trick is one of breaking a problem up into smaller pieces, then solving each of them separately. It is similar to the idea that Savitch used to prove his famous theorem relating nondeterministic and deterministic space complexity [119]. It is also related to the method used to get a "better" exponential time algorithm for the knapsack problem and for other problems.

Suppose that you want to see if there is a path from s to t in some graph. Savitch's trick is to guess a mid-point m and check:

$$s \rightarrow m \rightarrow t.$$

The trick that we use to prove these various lower bounds is based on:

$$s \rightarrow m_1 \rightarrow m_2 \rightarrow \ldots m_{k-1} \rightarrow m_k.$$

Some like to call the trick the speedup lemma. The point is that if you have some type of nondeterminism you *need not* check each of these steps, instead you can guess one and only check that one.

Here is another way to think about this idea. Suppose that you need to check the truth of a statement S. The statement may be very hard to check, and the obvious way of checking may be too slow. If you can find a set of statements $S_1, \ldots, S_k$ so that

$$S = S_1 \wedge \cdots \wedge S_k$$

then you can check each individual statement S_i separately. The problem is that this may still take too much time. Since space is "re-useable," for Savitch this works well:

$$\text{Space}(S) = \max_i \text{Space}(S_i).$$

However, we are interested in the case of time; thus, checking all statements S_i still takes too long,

$$\text{Time}(S) = \sum_{i=1}^{k} \text{Time}(S_i).$$

We need to get the maximum of the times to get a "too fast computation." The trick is then to use nondeterminism, which will allow us to only check one of the statements.

An important point about this trick. We want k, the number of sub-problems, to be as large as possible, since we only have to check one. However, the difficulty with making k large is that you need to "write down" in some sense the sub-problems. This tends to force k to be small. Roughly, this why the methods get stuck and cannot seem to prove arbitrarily large $\lambda's$.

This idea has a long history. It was certainly used by Ravi Kannan in complexity theory in his work on space limited computation [74]. But it was also used long ago to solve, for example, the discrete logarithm problem. Finding an x so that

$$a^x \equiv b \bmod p$$

can be done in time roughly $\sqrt{p}$ using the same method as above. The idea is to think of $x = cm + d$ where $m = \lceil \sqrt{p} \rceil$. Note, both c and d are bounded by m. We then rearrange and get the following equation,

$$a^{cm} \equiv b a^{-d} \bmod p.$$

We finally compute all possible left and right hand sides, and check for two that are the same modulo p.

14.4 Open Problems

An obvious goal would be to prove at least that $\lambda \geq 2$ is possible. There is strong evidence from Williams that this will require a new idea. He is able to show that in the current framework there is no proof that will reach this goal.

I am generally worried about any "impossibility proof." By this I mean a proof that says you cannot do something by a certain approach–Williams' result is of this type. Historically such proofs have been defeated, in other parts of mathematics, by subtle but significant shifts. However, Williams' proof does show that it will be impossible, without some new ingredient, to improve the SAT lower bounds. Clearly, we new a new idea.

I must admit I have thought of many ideas, none of which have lead anywhere. I will end this my telling you some of the dead ends that I tried.

- Try to select many sub-problems, but name them in some indirect manner.
- Try to use PCP technology. This is one of the deepest methods that we have concerning SAT, but there seems, to me, no way to exploit it.
- Try to break the computations up in a more arithmetical manner. We know that these methods are very powerful: Lance's work on the permanent is a prime example. Can we use the fact that the computations are coded as polynomials to improve these lower bounds? I tried, but have made no progress to date.

Good luck, I hope you see something that we have missed.

14.5 Notes

This appeared as sat-is-not-too-easy in the blog. Timothy Chow pointed to another famous mistake of Henri Lebesgue, who incorrectly asserted the projection of a Borel set is Borel.

Chapter 15
Ramsey's Theorem and NP

Endre Szemerédi is one of the greatest combinatorists in the world. He is famous, of course, for solving the Erdős-Turán arithmetic progression conjecture, proving the powerful regularity lemma, and for a many other terrific results. He has won many prizes, including the Pólya prize, Steele Prize, and the Schock Prize.

I shall talk about a potential relationship between combinatorial theorems and proof systems for tautologies. As you know I am not afraid to appear foolish, while suggesting approaches to problems that most think are false. In this post, it is an approach to obtain quasi-polynomial sized proofs for tautologies. I thought for a while that this could potentially succeed, but now I am less confident that it is on the right track. However, I would like to share it with you, and let you decide for yourself.

Since Szemerédi plays a role in my proposed proof system, I thought I would say something about him first. Years ago I looked at Szemerédi's paper that proves his famous arithmetic progression theorem. I do not claim to have understood the proof, but I wanted to see the masterpiece for myself.

An aside: I think we all can benefit from at least looking at some the great masterpieces. While the details may not be easy to follow, especially for a non-expert, just looking at these papers is an experience that is valuable. Take a look at Walter Feit and John Thompson's paper, Andrew Wiles's paper, Ben Green and Terence Tao's paper, the proof of the PCP theorem, and so on. Really.

Endre's paper was so complex that I recall it has a figure that is a flow-chart, which shows the dependencies of the various parts of the proof. However, the coolest part of the paper is–in my opinion– in the acknowledgment, where Szemerédi said:

I would like to thank Ron Graham for writing the paper.

Amazing. It was all Endre's work, but Ron wrote the paper.

When I was at Princeton working with Dan Boneh, we once needed a constructive version of Andy Yao's XOR lemma. At the time the proofs of this lemma were not strong enough for our application. So for a COLT paper we proved what we needed from scratch–luckily we only needed a special case, so the proof was not too difficult.

However, Dan and I still wanted to be able to prove the full case, and after getting nowhere we asked Szemerédi for help. He was kind enough to come down from Rutgers and visit me at my Princeton office to discuss the problem. After a while, he understood the problem and said to me that either he could solve it in the next day or two or the problem was hopeless.

Two days later I got a phone call from Endre, who said that he had proved what we wanted. I was thrilled and immediately responded that I would be happy to come up to Rutgers to hear the details. Endre said no, that would not be necessary, since the proof was "so easy." In the next few minutes he spoke fast and outlined a proof using an entropy type argument. I furiously wrote down all I could follow. Then, he said good-bye.

I remember running down to Dan's office and excitedly telling him we had the solution. The trouble was that my notes were meager and we were not Endre. But, in a mere three hours or so, we re-constructed his argument. As I recall, Dan and I wrote up a draft of the result with all three as authors, but we never tried to publish it. Does that make our Szemerédi number $i = \sqrt{-1}$?

15.1 Ramsey Theory and PCP

The famous Ramsey theorem will play an important role in the suggested proof system. For a graph G let as usual $\omega(G)$ be the size of the largest clique of G. Also define $R(G,k) \geq r$, if in every k coloring of the edges of G, there is a set of vertices S of size at least r, so that S forms a clique of G and all the edges in S are the same color. The famous Ramsey theorem states that for k fixed, $R(K_n,k)$ tends to infinity as n tends to infinity.

There are other Ramsey style theorems that are also needed. The following is a generalization of the above to "hypergraphs". These concern coloring of not edges, but coloring of l-cliques of the graph. Define $R(G,k,l) \geq r$, if in every k coloring of the l-cliques of G, there is a set of vertices S of size at least r, so that S forms a clique of G, so that all l-subsets of S have the same color. Also for fixed k and l, $R(K_n,k,l)$ tends to infinity as n tends to infinity.

Another important ingredient, for our approach is the PCP gap theorem for cliques. One of the best bounds is due to David Zuckerman [139]:

Theorem 15.1. *It is NP-hard to approximate the max-clique problem to within a factor of $n^{1-\varepsilon}$.*

Actually, under stronger assumptions ε can tend to 0 slowly.

15.2 A Proof System

Suppose that G is a graph that either has only *small* cliques all of size at most n^ε, or has at least one *large* clique of size $n^{1-\varepsilon}$. We do not know how in polynomial time to determine whether or not a graph is in either of these categories: small or large. The goal here is to present a simple proof system based on Ramsey theory that *might* be able to supply a quasi-polynomial size proof that G does not have a large clique.
   This is the proof system:

---

1. Let $r = R(K_m, 2)$ where $m = \lfloor n^{1-\varepsilon} \rfloor$;
2. Guess a 2-coloring of the edges of G;
3. For each set of r vertices of G, check if they form a mono-chromatic clique;
4. If all fail to be mono-chromatic, then state that G has no large clique.

There are several issues with this "proof system." Let's skip the discussion of step (1), and return to that later on. The size of the proof is dominated by the cost of step (3): this takes at most quasi-polynomial time, since r is at most order $\log n$. Next the proof system is *sound*: if it states that G has no large clique, then this is true. Note, if G has a large clique, then just that part of the graph will always have a mono-chromatic clique of size at least r. So the system is sound.
   The central question is simple: is the proof system *complete*? The only way that it could fail to be complete is if there were a graph G so that $\omega(G) \leq n^\varepsilon$, and yet for all 2-colorings, there is a mono-chromatic clique of size r. If there is no such graph, then the proof system is complete. Thus, a key question, that I do not know the answer to, is: Do such graphs exist?
   There is one loose end. In step (1) we need the value $r = R(K_m, 2)$. We fix this by allowing the proof system $\log\log n + O(1)$ bits of advice.

15.3 Another Proof System

We can also extend the proof system via hypergraph Ramsey theory. Again suppose that G is a graph that either has only *small* cliques all of size at most n^ε, or has at least one *large* clique of size $n^{1-\varepsilon}$.
   This is the second proof system:

---

1. Let $r = R(K_m, k, l)$ where $m = \lfloor n^{1-\varepsilon} \rfloor$;
2. Guess a k-coloring of the l-cliques of G;
3. For each set S of r vertices of G, check that S is a clique of G, so that all the l subsets of S are labeled with the same color.
4. If all fail this test, then state that G has no large clique.

As before, step (1) will be done through advice bits. The size of the proof is dominated by the cost of step (3): this takes at most quasi-polynomial time, since r and l are at most order $\log n$. Next the proof system is sound: if it states that G has no large clique, then this is true. Note, if G has a large clique, then just that part of the graph will always have a mono-chromatic clique of size at least r. So the system is sound.

Again the completeness is the open problem. The role of Szemerédi was to be an informal oracle. I ran similar ideas by him, years ago, and he seemed to think that it was not obvious whether or not the proof system worked. Any mistake is mine, but I thought that his comments were at least comforting.

The reason for the second system is that the first might be incomplete, while the second is complete. The advantage of the second system is our lack of knowledge of the behavior of these Ramsey numbers. The real reason is a hedge against the possibility the first system is incomplete. At the moment I have no strong intuition why the second system may be complete, mainly due to our poor knowledge about the behavior of generalized Ramsey numbers.

15.4 Open Problems

Of course the obvious question is to show that these proof systems are incomplete, especially if you believe that tautologies are hard. Or try to show that one of them is complete. Note, in the second proof system is really a family of systems, since we are free to select k and l.

Another possibility is to create other similar proof systems, perhaps based on other Ramsey type theorems. Other examples, would, in my opinion, be exciting. They would relate deep mathematical theory with our basic questions of complexity theory.

I think using the PCP theorem as part of a proof system seems to be a powerful idea. Again, while the proof system outlined is likely not to be complete, its incompleteness would shed light on Ramsey theory itself.

15.5 Notes

This appeared as a-proof-system-based-on-ramsey-theory in the blog. There was a lively discussion by Timothy Gowers, Luca Trevisan, Elad Verbin, and others on this post. Gowers had a kind comment, but believes that the proof systems I presented are likely to be incomplete. I think he is probably right, but I do not know if anyone has made any progress on these systems.

Chapter 16
Can They Do That?

Alan Turing is of course famous for his machine based notion of what computable means—what we now call, in his honor, Turing Machines. His model is the foundation on which all modern complexity theory rests; it is hard to imagine what theory would be like without his beautiful model. His proof that the Halting Problem is impossible for his machines is compelling precisely because his model of computation is so clearly "right." Turing did many other things: his code-breaking work helped win the war; his work on AI is still used today; his ideas on game playing, biology, and the Riemann Hypothesis were ground-breaking.

I plan to talk about the role of negative results in science, which are results that say "you cannot do that." I claim that there are very few meaningful negative results; even Turing's result must be viewed with care.

You may know that Turing took his own life after being treated inhumanely by his own government. Just recently the British Prime Minister, Gordon Brown, said that the treatment of Alan Turing was "appalling." Part of Brown's statement is:

Turing was a quite brilliant mathematician, most famous for his work on breaking the German Enigma codes. It is no exaggeration to say that, without his outstanding contribution, the history of World War Two could well have been very different. He truly was one of those individuals we can point to whose unique contribution helped to turn the tide of war. The debt of gratitude he is owed makes it all the more horrifying, therefore, that he was treated so inhumanely. In 1952, he was convicted of 'gross indecency' - in effect, tried for being gay. His sentence - and he was faced with the miserable choice of this or prison - was chemical castration by a series of injections of female hormones. He took his own life just two years later. ...

So on behalf of the British government, and all those who live freely thanks to Alan's work I am very proud to say: we're sorry, you deserved so much better.

The terrible treatment of one of the greatest mathematicians that ever lived cannot be undone, but we will continue to honor him every time we use his intellectual legacy.

I have an interesting story about Turing. The story—I believe—shows something interesting about Turing's character. I hope that the story is still appropriate to present—I think it still is, and I hope that you agree.

There are many stories about Turing—one of my favorites took place right after the start of World War II. At the time, Turing was already working at Bletchley Park. This work was fundamental to the war effort and eventually helped to win the war against Germany—especially fighting U-boats in the North Atlantic.

Right after the fall of France in May-June 1940, it seemed that Germany might invade England at any moment. Turing wanted to be ready to fight, but he had no idea how to fire a rifle, so he signed up for the Home Guard. This was a part-time way to join the army and learn the basics of firearms.

When he joined, the form that he had to sign asked the question: 'You understand by signing this form that you could be drafted into the regular army at any moment.' He signed the form, but he answered the question "No." An official apparently filed it, without bothering to read it. Turing then went to enough meetings to learn how to fire a rifle. When it became clear that England was not going to be invaded, he stopped going to the meetings.

The Home Guard officer for his locale later decided to call up Turing into the regular army. Little did the officer know that there was no way that this could ever have happened: since the secret work Turing was doing was critical to the war effort, Prime Minister Winston Churchill would have stopped it personally.

Turing went to see the officer anyway. At their meeting, the officer pointed out that Turing had signed a form that allowed Turing to be put directly into the army. Turing smiled and said, 'Take a look at my form.' There, where it said, You understand by signing this form that you could be drafted into the regular army at any moment, was Turing's answer "No." Apparently, Turing was thrown out of the meeting.

Let's now go on to discuss impossibility results, including the Halting Problem.

16.1 Can They Do That?

I love the "Far Side" cartoons by Gary Larson. One of my favorites is a cartoon of a western scene circa mid-1800's. Cowboys are in a wooden fort shooting at the attacking Indians. The Indians do not have rifles, like the cowboys, but they have bows and arrows and are shooting burning arrows at the cowboys. Some have hit the wooden fort, which is starting to burn. One cowboy says to another, "Are they allowed to do that?"

In my opinion this is crux of why an impossibility proof (IP) can be misleading. Of course the Indians can shoot burning arrows or any kind of arrow they wish, even if the cowboy's wooden fort is not prepared for them.

Thus, to me the key issue with any IP is whether the rules of what is or is not allowed are *complete*. The IP is only as strong as the assumptions of what the "attackers" can do. If the assumptions are too weak, then the IP results can be meaningless.

> Impossibilities are merely things of which we have not learned, or which we do not wish to happen.
> Charles Chesnutt

I like the latter part of this quote: "which we do not wish to happen." I think many IP's have that flavor.

16.2 Burning Arrows

Over the years many IP's were later shown to be not so impossible—that is, they were *wrong*. I would like to point out a few of my personal favorites. They all failed because their assumptions were incorrect: they did not allow for "burning arrows."

• **Halting Problem:** One of the great IP's is the impossibility of deciding whether or not a program will halt. Of course, this is due to Turing, and seems to be a very strong result. Not only is the halting problem impossible, but by Rice's Theorem any property of a program that depends on the program's behavior is impossible to decide in general. This theorem is due to Henry Rice.

Yet most "real" programs are written in such a way that it is often easy to see that they halt. If a program uses only *for* loops, for example, with fixed bounds, then it must halt. For many other classes of programs also it is not hard to see that they will halt.

So is the halting problem really impossible? Yes it is. However, that does not mean that the halting problem is impossible for programs that arise in practice. For example, Byron Cook of Microsoft has worked on a system called TERMINATOR [38]. This project has had good success in discovering whether a program can potentially hang the operating system.

Is the halting problem impossible: yes in general, but perhaps not in most examples that we care about. Perhaps.

• **Trisecting Angles:** Pierre Wantzel proved in 1837 that the ancient Greek problem of trisecting a general angle with a straight-edge and a compass is impossible. The proof shows that the angle that arises for some trisections is an angle that cannot be constructed with these limited tools. It is a rock-solid IP, but as Underwood Dudley points out in his wonderful book, the proof has not stopped many amateurs from trying to find constructions that circumvent the proof [44].

Trisection is solvable with slightly different tools, which are not much more complex than the allowed ruler and compass. If the ruler has two marks on it, then there is a solution. Or given a piece of string, one can trisect any angle. Again Wantzel's proof is an IP that is correct, but it does not completely close all the doors on the problem. A marked ruler can play the role of the burning arrows.

• **Airplanes:** William Thomson, Lord Kelvin, is famously quoted as saying, "I can state flatly that heavier-than-air flying machines are impossible." However, there

is some evidence he did not say this, and even if he did I doubt that he had a real IP.
So I will not count this as a flawed IP.

• **Rockets:** There was a paper published in the early 1900's on the possibility
of a chemical rocket ever getting into orbit. The paper gave a careful and tight IP
that showed this was impossible. Of course, today, there are countless satellites in
orbit around the Earth, so where was the error in the IP? The paper used careful and
correct laws of physics and reasonable assumptions about the amount of energy that
could be obtained from chemical reactions.

The mistake was simple: the proof assumed implicitly that the rocket had constant mass. Clearly, as a rocket burns fuel it gets lighter, and this was the mistake
that killed the IP. Note, there was a real insight in the flawed IP—the motivation for
why rockets usually have many stages. Rockets stages allow mass to be shed at a
quicker rate than just losing the mass of the fuel that is burnt. Early rockets used
three or even four stages to place their payload into orbit, while today it is possible
to get into orbit with two stages—one might argue that the space shuttle uses 1.5
stages.

Along similar lines, when physicist Robert Goddard published his groundbreak-
ing paper "A Method of Reaching Extreme Altitudes," [56], the New York Times
reacted sharply, by heaping severe criticism on his ideas, even attacking his integrity.
The basis of their attacks was the same as above, implicitly assuming that a fixed
mass system cannot accelerate in a vacuum.

After a gap of nearly 50 years, three days before man set his foot on the moon,
the NYT eventually came around and rescinded their stance — they published a
short correction article to their 1920 article.

• **Crypto Protocols:** Cryptographic-protocols are one of great sources of IP's
that are often later shown to be flawed. The reason is simple: the IP rests on assumptions about what the attacker can do. In cryptography the attacker is often well
motivated, endowed with large resources, and very clever. Thus, the IP's are "broken" not by direct attacks, but by showing that the IP's assumptions are flawed.

A famous one is Johan Håstad's attack on RSA, which used the Chinese Remainder Theorem (CRT) —one of my favorite theorems—see Chp.5. He showed
that sending the same message via many RSA systems was dangerous: the message
could be found without directly breaking RSA [19]. I was involved in another attack that also used the CRT to break crypto-systems such as RSA in the presence of
faults. This was joint work with Dan Boneh and Rich DeMillo [20].

More recent examples are problems with crypto-protocols that are composed
with other protocols, and also protocols that encrypt messages that are used later
as keys in another system. The latter are called *circular security* models, and there
is active research into their properties.

• **Quantum Protocols:** Charles Bennett and Gilles Brassard in 1984 suggested
how to use quantum mechanics to build a key exchange system that would be impossible to break. They had an IP that was not based on a complexity assumption,
but on the correctness of quantum mechanics. This was a brilliant idea then, and
is still one of the great insights about practical applications of quantum methods to
security.

However, the exact proof that the system, BB84, is secure is not quite right. There are a number of attacks that were missed in the proof. For example, some sources of single photons sometimes put out multiple photons. This allows the photon number *splitting attack* to be applied [17]. Again these were flaws in the assumptions of what an attacker could do.

16.3 Impossibility Results

So what is the role of IP's? Are they ever useful? I would say that they are useful, and that they can add to our understanding of a problem. At a minimum they show us where to attack the problem in question.

If you prove that no X can solve some problem Y, then the proper view is that I should look carefully at methods that lie outside X. I should not give up. I should look carefully—perhaps more carefully than is usually done—to see if X really captures all the possible attacks that are possible on problem Y.

What troubles me about IP's is that they often are not very careful about X. They often rely on testimonial, anecdotal evidence, or personal experience to convince one that X is complete.

16.4 Open Problems

The open problem is to please view all IP's with a certain degree of doubt—question, question, and question the assumptions. Considering, for example, "Natural Proofs" as an IP raises the question: How strong is it? Are there ways around it? See the discussion in Chp. 11.

I of course am open to P=NP as a possibility. But even a proof that P$\neq$NP viewed as an IP should be taken as a direction, not as an absolute statement. I have talked before about the many ways that P$\neq$NP can be true, please think hard about these.

16.5 Notes

This appeared as are-impossibility-proofs-possible in the blog. Many added their own IP results. Amit Chakrabarti said his favorite was the impossibility of solving a general quintic by radicals. Jonathan Katz talked about airplanes. Thanks again to all who took time to make a comment. Thanks to Martin Schwarz, John Sidles, Tyson Williams, and everyone else.

Chapter 17
Rabin Flips a Coin

Michael Rabin is one of the greatest theoreticians in the world. He is a Turing Award winner, with Dana Scott in 1976, a winner of the Paris Kanellakis Award, a winner of the EMET Prize, and many others. What is so impressive about Rabin's work is the unique combination of depth and breadth; the unique combination of solving hard open problems and the creation of entire new fields. I can think of few who have done anything close to this.

For example, one of his results is the famous proof, in 1969, that the monadic second-order theory of two successors is decidable–$(S2S)$ [45]. I will not explain this here, but trust me this is is one of the deepest decidability results in logic. What is even more remarkable about this theorem is the myriad number of corollaries that flow from it. He also helped create the field of modern cryptography, and has made contribution to almost all parts of computer science from distributed computing to secure operating systems. However, one of my personal favorite of his results is the randomized linear time pattern matching algorithm he created in 1987 with Dick Karp [78]. This work is elegant and simple: it is based on just doing the "obvious" brute force algorithm with a dollop of randomness to make it run in linear time. What a wonderful result.

I first met Rabin when I was a graduate student at, Carnegie Mellon, probably around 1970. One day Rabin visited CMU and my advisor at the time–Don Loveland–set up a meeting between the three of us. Michael kindly asked me what I was working on at the time. Then, I was working on a problem that concerned the power of a certain type of tree automata. Rabin listened to my problem and asked if it would be okay if he thought about the question. I replied of course–I wanted to know the answer more than anything.

A few days later I was entering a classroom where the theory qualifier was being given. Loveland was proctoring the exam and causally told me that Rabin had just called him with a new idea on how to solve my problem. Don said it was unclear if it would work, but he would tell me the idea later on, *after the exam*. I still remember sitting down at my desk and consciously thinking: I could do the exam and likely pass, or I could try to figure out what Rabin's new idea was and surely fail. But I could not do both. I decided to work on the exam–not an easy choice. I found out the next day, Rabin's idea was very clever, but was not enough to crack the problem. It is still open today.

Since then I have had the pleasure to work with Michael and to count him as a close colleague and friend. I will talk about his work that essentially brought "randomness" to the design of algorithms.

17.1 The Nearest Neighborhood Problem

Rabin my not have been the first to use randomness in algorithms, but he was the last to introduce the notion. Often in mathematics the key is not who discover something first, but who discovers it last. Rabin discovered the power of randomness last.

He used randomness to solve a geometric problem about points in the plane. His result was that with randomness a certain problem could be solved in linear time— the previous best was quadratic time. He gave a series of talks on his ideas that quickly spread the word throughout the community that randomness is powerful. In a short time, there were many other papers using randomness. Today of course randomness is a key tool in the design of algorithms. We teach whole courses on it, and there are whole conferences just on this critical topic. It such a part of theory that it may be hard to believe that there was a time when algorithms were always deterministic.

The problem was the *nearest neighbor* problem for the plane. Let n points

$$x_1, x_2, \ldots, x_n$$

be in the unit square. The problem is to find the two points x_i and x_j that are the closest in Euclidean distance. It is convenient to assume that there is a unique closest pair. If there are several with the same minimum distance, then Rabin's algorithm still works. However, this assumption just makes the discussion easier to follow.

17.2 Rabin's Algorithm

Here is the outline of Rabin's algorithm. Clearly, the problem can be solved in time $O(n^2)$ by a brute force method. Just compute the distance between each pair of points. Rabin had the clever idea of randomly selecting $\sqrt{n}$ of the points: call them S. Then he computes the minimum distance between the points in S. This he does by the brute force algorithm, but now clearly this takes only

$$O((\sqrt{n})^2) = O(n)$$

time. Call d the minimum distance among the points in the sample set S.

He then proceeds to imagine that the unit square is divided up into a "checkerboard" where each small square is size d by d. Clearly, there may be a huge number of these squares. If, for example, d is very small then there could be many more squares that points. In any event no algorithm that hopes to run in linear time can afford to "look" at all the squares. So we must figure out a way to avoid that. Let's call a square *filled* if it contains one or more poins from the n original points. Note, there are at most n filled squares. Rabin then does the following: First, for every filled square he computes the closest pair within the square. This is again done by brute force. Second, for every pair of adjacent filled squares he computes the closest pair between their points: call squares adjacent if they share even one point. He then claims that he has found the closest pair.

This is really quite a simple argument. No matter what d is the closest pair must be either in a square together or must be in touching squares. If they were separated by a square, then their distance would be at least d.

There are two final pieces of his algorithm. First, we must show that we can find the squares that are filled fast. As pointed out earlier we cannot afford to look at all the squares. This is done by a hashing technique, and was the subject of some discussion at the time. Clearly one needs a model of computation that allows the finding of the filled squares fast. Since this is not a random idea we will not comment further on it other than to say it is an issue. See Steve Fortune and John Hopcroft for a discussion about this issue [51].

What is much more interesting is the running time of the algorithm. Let f_k be the number of points in the kth filled square. It is easy to show that the cost of the comparing of the distances in the filled squares and in the adjacent squares is bounded above by

$$C = \sum_k O(f_k^2).$$

Note, the value of this sum must be linear for Rabin's algorithm to run in linear time. The value C depends on the quality of the sampled d. If d is too big, then the sum can easily be super-linear. The tour-de-force is that Rabin is able with difficult analysis to show since the sample had $\sqrt{n}$ points the expected size of C is linear in n.

17.3 A Slight Variation on Rabin's Theme

After Rabin visited me at Yale and gave his talk on this nearest neighborhood algorithm we tried to reproduce the proof that C is linear. The proof is difficult, and we could not reconstruct it. The trouble is that the sample S is of size $\sqrt{n}$ and thus has about n distances. If these were independent they would give a good estimate of the real minimum distance. However, they clearly are not independent. If a, b, c are points in S, then if a, b are close and b, c are far apart, it seems likely that a, c would also be far apart. This is what makes the analysis so hard.

We soon realized that there was a way to fix Rabin's algorithm so the analysis would be simple. The key is not to sample points, but rather to sample distances. More precisely, select n pairs of points and compute the distances. Let d now be the minimum among these distances. It is now quite simple to prove that C is linear in n. The reason it is so much easier is that now the distances are all mutually independent by definition.

17.4 Open Problems

Rabin's algorithm was not as important as the demonstration of the power of randomized algorithms. The problem of finding nearest neighbors in higher dimension is still a problem that is open.

17.5 Notes

This appeared as rabin-flips-a-coin in the blog. Samir Khuller pointed out related work he did with Yossi Matias on the nearest neighbor problem [81].

Chapter 18
A Proof We All Missed

Neil Immerman and Robert Szelepcsényi are two computer scientists who are famous for solving a long standing open problem [69, 126]. What is interesting is that they solved the problem at the same time, in 1987, and with essentially the same method. Independently. Theory is like that sometimes.

They solved the so called "LBA problem." This problem was first raised in the 1960's. When I was a graduate student we all knew about the problem. But no one thought about the question as far as I could tell. It stood unsolved for so long for two reasons, in my opinion. First, no one really worked on the problem. It appeared to be unapproachable. Second, we all guessed wrong. The conventional wisdom was that the problem went one way. That was obvious. But Immermann and Szelepcsényi showed that it went the other way. Sound familiar? Could P=NP go the same way? Who knows?

There is one more comment I must add. Robert was a student when he solved the problem. The legend is that he was given a list of homework problems. Since he missed class he did not know that the last problem of his homework was the famous unsolved LBA question. He turned in a solution to the homework that solved *all* the problems. I cannot imagine what the instructor thought when he saw the solution. Note, it is rumored that this has happened before in mathematics. Some believe this is how Green's Theorem was first solved. In 1854 Stoke included the "theorem" on an examination. Perhaps we should put P=NP on theory exams and hope . . .

18.1 The LBA Problem

The LBA problem was first stated by S.Y. Kuroda in 1964 [86]. The problem arose in a natural way since in the 1960's there was great interest in the power of various types of grammars. Grammars were used to describe languages. There was a hierarchy of more and more complex types of grammars. At the bottom were grammars for regular languages, next came context-free grammars. Then, came context-sensitive grammars which were very powerful. They could describe quite complex languages.

A natural question that arose immediately is: if L is a context-sensitive language is its complement $\bar{L}$ also one? This was easy to resolve for the regular case, *yes*, and for the context-free case, *no*. For the context-sensitive case the answer was elusive. The conventional wisdom was that it had to be no, but no one seemed to have a clue how to proceed.

A language is context-sensitive precisely if it is accepted by a nondeterministic Turing machine whose work tape is exactly the same length as the input. Such a machine is called a *Linear Bounded Acceptor* or LBA. The LBA problem thus asked, whether or not such machines were closed under complement. Since LBA's are nondeterministic, the conventional wisdom was simple: there must be a context-sensitive language whose complement is not context-sensitive. Note, the intuition was reasonable: how could a non-deterministic Turing Machine with only linear space check that *no* computation accepted some input? The obvious methods of checking all seemed to require much more than linear space.

18.2 The Solution

George Pólya suggests that when trying to prove something, often a good strategy is to try to prove more. This and other great suggestions are in his classic book: *How to Solve It* [114]. The rationale behind his suggestion is simple: when you are proving "more" you often have a stronger inductive statement.

Neil and Robert did exactly this. They did not determine whether or not the start state can reach the final state. Instead they counted exactly how many states could be reached from the start. Counting how many states are reachable gives them a "Pólya" type advantage that makes their whole method work.

The key to their solution is the following cool idea. Suppose that we have any set S of states. Call a nondeterministic machine G a *weak generator* of S provided it always generates a sequence $s_1, \ldots, s_t, v$ where each s_i is a state and $v \in \{0, 1\}$. If $v = 1$, then the states s_i are all distinct and $S = \{s_1, \ldots, s_t\}$. Further, the machine has some nondeterministic computation that outputs a sequence that ends in 1.

Let S_k be all the states that are reachable in at most k steps from the start state. Reducing the LBA problem to calculating the size of S_k for a large enough k is an easy remark. See their papers for this [69, 126]. Neil and Robert's great insight is that there is a weak generator for S_k for any k. Moreover, this generator can be constructed from *just* the cardinality of S_k. This is the remarkable idea: knowing only the *size* of the set S_k one can construct a weak generator for S_k.

We now need to explain how they can create a weak generator for S_k from just the cardinality of S_k. Let $m = 0$. The idea is to go through all states s in a fixed order. For each s we will guess whether s is in S_k. If we guess no, then we move on to the next state. If we guess yes, then we guess a path from the start to s of length at most k. If the guessed path is correct, then we output s and increment the counter m by 1. If the path is incorrect we move onto the next state. When we have gone through all states s we check the value of m. If it is equal to the cardinality of the set S_k, then we output $v = 1$. Otherwise, we output $v = 0$.

The key is this is a weak generator for S_k. All the properties are easy to check. The main point is that if the machine guesses wrong during the computation, then some state will be missed and m will not equal the size of the set S_k. Very neat.

The last point is that we can construct the size of S_k inductively. Clearly, the size of S_0 is 1. Suppose that we know the size of S_k. We will show how to construct the size of S_{k+1}. Since we have the size of S_k we have a weak generator G for S_k. We will use that to compute the size of the set S_{k+1}. Let $m = 0$ again be a counter. We again go through all states s in some fixed order. For each state s we run the generator G. We can assume that the generator creates $s_1, \ldots, s_t, 1$. If the last is not 1, then we just reject later when we see this. Otherwise, for each s_i, we check (i) is $s = s_i$ or (ii) is s reachable in one step from s_i. In either case we increment m by 1 and move on to the next state s'. If both fail, them we try s_{i+1}. It is not hard to see that this computes the number of states that are reachable from the start in $k + 1$ or fewer steps.

18.3 A Proof From the Sixties

With all due respect to Neil and Robert there is nothing about their clever solution that could not have been done 20 years ago. They solved the LBA problem directly, and they used no fancy methods or techniques that were developed during the intervening years. It's very clever, but there is nothing there that was unknown in the sixties. I often wonder whether there is some similar simple idea that would resolve P=NP? Is there some basic idea(s) that we are all missing? I hope that someone reading this will see something that we all have missed. That would be great.

18.4 Open Problems

An obvious question is: can any similar method be used to show that NP is closed under complement? One interesting note is that for their counting method to work they do not need to get the exact cardinality of the sets S_k. Even a coarse approximation would suffice. This suggests some ideas that might work in other contexts.

18.5 Notes

This appeared as we-all-guessed-wrong in the blog. The most interesting comment was from Neil, himself, who stated the story of how Robert solved the problem was not true. Oh well.

Chapter 19
Barrington Gets Simple

David Barrington is famous for solving a long standing open conjecture. What I love about his result is that not only was it open for years, not only is his solution beautiful, but he also proved it was false. Most of us who had worked on the problem guessed the other way: we thought for sure it was true. Again conventional wisdom was wrong. This makes me wonder about other problems.

19.1 Bounded Width Computation

There are many restricted models of computations. Without these weak models we would have no lower bounds. Proving in a general model–such as unrestricted boolean circuits–that an explicit problem cannot be computed efficiently seems to be completely hopeless. However, once some restriction is placed on the model there is at least *hope* that an explicit problem has a lower bound. Barrington worked on a restricted model called *bounded width* computations. While most (many? all?) of the community believed that this model was extremely weak and would yield lower bounds, he shocked us and showed that bounded width computations were quite powerful.

A bounded width computation is a way of computing a boolean function. Suppose that $x_1, x_2, \ldots, x_n$ are the inputs to the function. You can think of a bounded width computation as a line of boxes. Each box is associated with some input x_i. The box gets bits from the left, and uses these and the input x_i to decide on its output which goes into the box on its right. Of course the first (leftmost) box is special and gets a fixed value from the left; the last (rightmost) box outputs the answer. The size of the computation is the number of boxes or equivalently the length of the line. So far anything can be computed by such devices. To see this, just have each box concatenate the left input with their x_i. Then the last box gets $x_1 x_2 \ldots x_n$ as a string and so can easily determine the value of any boolean function.

The key is that we restrict the number of bits that can be passed from one box to the next. In particular, bounded width computations must have their boxes only receive and send $O(1)$ bits *independent* of the number n. This is the restriction that makes the model interesting. While the bits that can be passed must stay bounded, the length can grow faster than n. Therefore, different boxes can use or read the same input x_i. Without this added ability the model would reduce to just a finite state type machine and would be quite weak when coupled with the bounded restriction.

The question that Barrington solved was exactly what can bounded width computations compute with size at most polynomial? The conventional wisdom at the time was that such a model could not compute, for example, the majority function. My intuition–that was dead wrong–was that in order to compute the majority function the boxes would have to somehow pass the *number* of 1 bits that are to their left. This of course does not work because this grows as $\log n$ and is certainly not constant. I even worked with Ashok Chandra and Merrick Furst and proved a non-linear lower bound on the size of a bounded width computation for majority [29]. While our bound was just barely non-linear, we thought that we might be able to improve it and eventually show that majority was hard. By the way this work with Chandra and Furst is discussed further in Chp. 38.

Barrington proved the following beautiful theorem [14]:

Theorem 19.1. *Any boolean formula of size S in can be computed by an $O(1)$ bounded width computation of size at most $S^{O(1)}$.*

The theorem shows immediately that we were wrong. Since majority can be shown to have polynomial size formulas his theorem showed that the conventional wisdom was wrong. Often a weaker theorem is stated that shows that if the formula is depth d, then the bounded width computation is at most size $2^{O(d)}$. The above is really not stronger since it has long been known that any formula of size S could be converted into one of $O(\log S)$ depth keeping the size polynomial. The corresponding question about depth for *circuits* is still wide open: we have no idea how to make a circuit have smaller depth and not explode its size.

The fact that formulas can be made to have logarithmic depth in their size without exploding their size has a long and interesting history. I do not have the space now to explain the main idea but the key is that binary trees have good recursive properties–let's leave it at that for now. Philip Spira worked on the first result, then Richard Brent, Vaughan Pratt, and others. See Sam Buss's paper for a recent result and some more background [21].

I will not give the full proof of Barrington's Theorem. I will, however, give you the key insight that he made. His brilliant insight was to *restrict* the values that boxes pass from one to another to values from a finite group G. Clearly, as long as the group is fixed this will satisfy the bounded width restriction: a group only takes $O(1)$ bits. What is not at all obvious is that by restricting his attention to finite groups he could see how to make the bounded width model compute any formula. I think this is a nice example of a trick that we should use more often. By adding constraints he made the problem easier to understand: Henry Kissinger once said, "The absence of alternatives clears the mind marvelously." By focusing on groups David was forced to use certain constructs from group theory that make his method work.

Barrington's theorem is proved by showing that every formula can be computed by a bounded width computation over the same fixed group G. The group is assumed to be a non-abelian simple group. A key notation is that of *commutator* of two group elements: the commutator of x,y is $[x,y] = xyx^{-1}y^{-1}$. Since G is non-abelian and simple it follows that every element is the product of commutators.

The way that a bounded width computation will compute a formula is simple: if the formula is true let the output be the identity element 1 of the group; if the formula is false, then the output will be a g that is not the identity. To make the induction work we will insist that g can be any non-identity element. This is a standard Pólya trick that makes the proof work nicely. Clearly, we can handle any input variable x_i. Negation is also simple: just multiple by g^{-1} and this flips the value of 1 and g.

The key case is how do we do an "or" gate? Suppose that we want to compute $\mathscr{A} \vee \mathscr{B}$. Since we want the values of this computation to be $1,g$, we select a,b so that $[a,b] = g$. This is possible. Let use A (resp. B) to denote the bounded width computation that computes $\mathscr{A}$ (resp. $\mathscr{B}$). Then, the claim is that

$$C = A \rightarrow B \rightarrow A^{-1} \rightarrow B^{-1}$$

computes the "or". It should be clear that if either A or B outputs 1, then C outputs 1. What happens if they both do not output 1. Then we get the value

$$g = aba^{-1}b^{-1}.$$

But this is the value that we needed. Pretty clever.

19.2 Open Problems

This area is rich with open problems. For starters we do not understand what groups are needed to make Barrington's Theorem work. He shows that any non-abelian simple group is enough. In the opposite direction it is known that no nilpotent group will work. The big open question is, what about solvable groups? We believe that they should not be able to compute anything of interest, but we all guessed wrong before so perhaps we are wrong again. It is not hard to see that the exact technology that David uses will not work for solvable groups. But that, of course proves nothing. I consider this one of the most interesting open questions in this area of theory. The reason that nilpotent groups can be handled is that they are always direct products of p-groups. And p–groups can be handled by known results.

I have a completely wild idea that I hesitate to share lest you stop reading my book and think I have lost it. But I cannot resist so I will tell you anyway. Suppose that you could prove from first principles that no group of odd order could compute the majority function. This would be a major result, since it would imply the famous (to group theorists at least) *Odd Order Theorem* [47] of Walter Feit and John Thompson. They showed that an odd order group is simple only if it is a cyclic group of prime order: thus, there are no interesting simple groups of odd order. So if no odd order group can compute majority, then by Barrington's Theorem no odd order group is simple. The original proof of the Feit-Thompson Theorem is over 200 journal pages–255 to be exact.

A possible, more doable idea would be to prove the (also famous) pq theorem of William Burnside. Burnside's theorem in group theory [58] states that if G is a finite group of order $p^r q^s$ then G is solvable. Hence each non-Abelian finite simple group has order divisible by three distinct primes.

The reason I think that there is at least some hope that complexity theory could prove such theorems is that we bring a completely different set of tools to the table. I think it could be the case that these tools could one day prove a pure theorem of mathematics in an entirely new manner. That would be really cool.

19.3 Notes

This appeared as barrington-gets-simple in the blog. I thank Andy Parrish for a correction. Also many thanks to David Barrington for some wonderful historical comments on his great discovery.

Chapter 20
Exponential Algorithms

Alan Perlis won the first Turing Award for his pioneering work on computer languages. He helped create the now famous department of computer science at Carnegie-Mellon University, then later in his career he created another department at Yale University. Alan was always smiling even though he spent much of his adult life in a wheel chair. He delighted in all things, especially anything about computing. You could walk into his office anytime and chat about just about anything. My connection with Alan was as a graduate student at Carnegie-Mellon, and later as a junior faculty at Yale when he was the chair there.

Perlis was famous for many "sayings". He talked about "one man's constant is another's variable," he coined the term "Turing Tar-pit." The latter referred, of course, to the fact that just about anything interesting that one wanted to do with computers was undecidable. He talked about the "IBM fog descending over all"– recall this was when IBM was *the* dominant player in computers.

20.1 Polynomial vs. Exponential Algorithms

Alan had one saying that is quite relevant to P=NP: he once said, "for every polynomial-time algorithm you have, there is an exponential algorithm that I would rather run." His point is simple: if your algorithm runs in n^4 than an algorithm that runs in $n2^{n/100}$ is faster for $n = 100$. While this is a simple point I think that it is a profound one. When we talk about P=NP we must not forget that the actually running time of the algorithm is the key. For example, if there is a factoring algorithm that runs in time $n2^{n/100}$ such an algorithm would easily break all today's factoring based crypto-systems. Yet its asymptotic performance would pale when compared to the best known factoring methods.

20.2 An Exponential Algorithm for Knapsack

One of my favorite algorithms is one for the knapsack problem that runs in time exponential in n, but the exponential term is $2^{n/2}$ instead of the obvious 2^n. This is a huge improvement over the obvious method.

Recall the knapsack problem is the question of whether or not given the positive integers $a_1, \ldots a_n$, there is a subset I of the indices 1 to n so that

$$\sum_{i \in I} a_i = b$$

where b is another given positive integer. The obvious algorithm would try all subsets I. Since there are $2^n - 1$ non-empty subsets the algorithm takes a long time.

The cool algorithm makes one small observation. The above equation can be re-written as follows:

$$\sum_{i \in I} a_i = b - \sum_{j \in J} a_j$$

where $I \subseteq \{1, \ldots, n/2\}$ and $J \subseteq \{n/2+1, \ldots, n\}$. (Yes, we assume that n is even, but if n is odd the method can easily be fixed.) Then, the algorithm computes two sets. The first consists of all the values of

$$\sum_{i \in I} a_i$$

where $I \subseteq \{1, \ldots, n/2\}$; and the second consists of all values of

$$b - \sum_{j \in J} a_j$$

where $J \subseteq \{n/2+1, \ldots, n\}$. We then simply check to see if these two have a value in common. The point is that they have a value in common if and only if the knapsack problem has a solution. The punch-line is that the cost for this is now dominated by the number of values in each set: $2^{n/2}$. Note, checking if they have a common value can be done either by hashing or by sorting. But, in either case, this takes time just slightly larger than the size of the sets. Very neat, I think.

20.3 Open Problems

The above algorithm relies on the following idea. Suppose that A and B are finite sets of integers, and define $A + B = \{a + b \mid a \in A \text{ and } b \in B\}$. Then, determining whether or not x is in $A + B$ can be done in time $O(|A| + |B|)$ if hashing is used. (If you use sorting the time goes up by a log-term.) Note, this is potentially much faster than a brute force search of the set $A + B$.

A natural problem is what happens with a set like $A + B + C$? Is there a way to tell if x is in this set in time $O(|A| + |B| + |C|)$, for example? If this is possible, then the knapsack problem has an algorithm whose exponential term is $2^{n/3}$. I have thought about this and related approaches quite often. I have yet to discover anything, I hope you have better luck.

20.4 Notes

This appeared as polynomial-vs-exponential-time in the blog. The proper citation for the knapsack result: it is due to Ellis Horowitz and Sartaj Sahni [67]. This was pointed out in comments on my blog by Cris Moore via his friend Stephan Mertens.

Chapter 21
An EXPSPACE Lower Bound

Albert Meyer, in the first decade of his career, was one of the top theorists in the world, yet you may not know his work in detail. The cause, most likely, is that years ago he closed down his research program in complexity theory, moving into the area of the logic and semantics of programming languages. While he was a complexity theorist, he helped invent the polynomial time hierarchy, he proved lower bounds on real-time Turing Machines, he showed huge lower bounds on various logical theories, he proved the sharpest nondeterministic time hierarchy theorem known, and more. Much of this was joint work with some of the best of the best: Mike Fischer, Mike Paterson, Charlie Rackoff, Larry Stockmeyer, and others.

I met Albert when he spent a year at Carnegie Mellon: he taught me basic recursion theory. I remember in class one day he mentioned the infamous *Collatz problem* on the mapping $x \to 3x + 1$: I spent several weeks trying to prove something–anything–about this problem with no success at all. Albert listened to my attempts, but I got nowhere.

Later, he moved to MIT, where he has been ever since. This chapter is about some joint and non-joint work that we did on a decision problem called the *Vector Addition Reachability Problem*. I have worked, over the years, on a number of decision problems; and have always liked them. There is something pleasing about such problems: upper bounds are fun because they show you can do something that previously was impossible, and lower bounds are fun because they show how hard something is.

I have to tell you a bit about Albert in order to explain the history of this work. Just a bit. Albert was always intense, he was quick, he was full of ideas, and he was sometimes brisk. Yet, I loved talking with him, listening to his talks, and working with him. The theory community–and me personally–were diminished when he left the field.

Once at a conference I started to tell Albert about my latest result. Right after I had stated the theorem and started to explain the proof, he interrupted me and said "be quiet." Clearly, he was trying to see if he could prove my result in real-time. Now the result was nice, not earth shattering, but I was pretty miffed that he would try to do this. So I complained to Albert that I should get credit for thinking of the problem, not just for the solution. He thought about that, and agreed. Along with his quick mind, he had a number of sayings that were very helpful. My favorite, which I still use, is his rule of 3:

Prove the theorem for 3 and then let 3 go to infinity.

The point, of course, in mathematics 2 "what-evers" is often special. But if you can prove the result for the case of 3, then you are often golden. Like any rule there are exceptions, but it is a good one to know.

21.1 Vector Addition Systems

For vectors from $\mathbb{Z}^d$, we will say that $v \geq 0$ or that v is non-negative provided, $v_i \geq 0$ for $i = 1, \ldots, d$. A d-dimensional *Vector Addition System*(VAS) consists of a non-negative vector s and a finite set of arbitrary vectors A where s and A are all in $\mathbb{Z}^d$.

The key notion is that of *reachable vectors*: the set of reachable vectors R of a VAS $\langle s, A \rangle$ is the smallest set of non-negative vectors so that $s \in R$, and if $x \in R$ and $a \in A$, then $x + a \in R$ *provided* $x + a$ is a non-negative vector.

Thus, the reachable vectors can be viewed in the following manner. You start with the special vector s. Then, you keep adding on vectors from A to the current vector–as long as you stay non-negative. Any vector that can be constructed in this manner is a reachable vector.

Without the non-negative constraint the set of vectors reachable is easy to describe. Suppose that $A = \{a, b, c\}$ as an example. Then, the set of reachable vectors is:

$$s + xa + yb + zc$$

where x, y, z are non-negative natural numbers. This is a very simple geometric set, and it is easy to determine what vectors are in this set. At worst it can be written as an integer programming problem; thus, it is clearly decidable. The key to the VAS notion of reachable is that the *order* that vectors from A are added to the current state is critical. This is what makes the notion of reachable difficult to understand.

Reachable captures a computational process, not just a geometric notion. This is why the behavior of VAS's is complex, and partly why the study of them is exciting. There are several additional reasons why we studied them. First, they were actually isomorphic to a popular model called Petri Nets. Thus, questions about Petri Nets could be converted into questions about VAS's.

Second, they were close to several other decision problems. Third, there is a close connection between VAS's and commutative semigroups–more on this later. Finally, the reason they were really interesting to me is that very strong theorists—Meyer, Paterson, Rabin–had worked on various aspects of their structure. For a me, a junior faculty, solving a problem near results they had worked on seemed like a good idea.

21.2 The Result

The obvious questions about VAS's are: is the set of reachable vectors finite? and can the system reach a particular vector? The first question is relatively easy to solve, although the best known running time is Ackermann like–the upper bound is astronomical. The second problem is the one I began to work on, and eventually I was able to prove:

Theorem 21.1. *The reachability problem for Vector Addition Systems is* EXPSPACE *hard.*

One curiosity is that I proved this before the following was proved:

Theorem 21.2. *The reachability problem for Vector Addition Systems is decidable.*

Clearly, lower bounds–my theorem–are a bit more interesting if the problem is decidable. My theorem would have been wiped out, if someone had been able to prove that the reachability problem was undecidable. I had thought hard about the upper bound, but failed completely, so I switched to proving a lower bound. You do what you can.

21.3 Getting The Result

The proof of the hardest result for VAS reachability started with a conversation with Meyer at a STOC conference. He told me that he had just proved, with Paterson, the n-dimensional VAS reachability problem required at least PSPACE. I had been thinking about VAS's and realized instantly that I could prove the same result with very many less dimensions. I knew a couple of tricks that Albert did not. I played it cool and said nothing to Albert. But once back at home–Yale at the time–I worked hard until I had a proof of the EXPSPACE lower bound.

I then called Albert to tell him the news. It was a strange phone call. He did not believe that I could prove such a theorem. The conversation was Albert listing a variety of mistakes that I must have made. I kept saying "no I did not do that", or "no I realize that". Finally, the call ended without me saying anything to him about the details of the proof.

As soon as I hung up, I realized there was a "bug" in my proof. Not anything that Albert had raised, but a bug nevertheless. I still remember talking a deep breath and saying to myself that I can fix this. In less than an hour the bug was gone and I had the lower bound. Months later I got to explain the details to Albert.

21.4 Main Ideas of The Proof

There are several ideas that used in the proof that reachability is EXPSPACE hard.
A M-counter is a device that can support the following operations:

1. The counter always holds a number in the range $0, \ldots, M$;
2. Set the counter to 0;
3. Increment the counter by 1;
4. Decrement the counter by 1, provided it is positive;

A standard result is that $O(1)$ M-counters can simulate a Turing tape of length L
provided
$$M \geq 2^{2^L}.$$

Thus, we need only show that a VAS of size n can simulate $O(1)$ counters of size

$$2^{2^n}.$$

The obvious problem is that we do not have counters. The main trick is to show
that we can use the VAS abilities to simulate real counters.

The way that I did this is I first showed that you could think of a VAS as a strange
kind of non-deterministic counter machine–call them *V machines*. A V machine had
a finite state control, and n counters. It was non-deterministic and could do only two
types of operations: it could add 1 to a counter or subtract 1 from a counter. However,
if it subtracted 1 from a counter that was currently 0, the computation would dies. It
also had no direct way of testing whether or not a counter was 0.

If I could add the ability to these V machines to test for 0, then they would be
able to compute anything. I realized that this was unlikely. I did figure out how to
make the machines able to test for 0 provided the numbers stored in the counters
never were larger than
$$M = 2^{2^n}.$$

This was enough for the lower bound.

The trick to make 0 testing possible was the following idea. Suppose that we had
a T-counter C, which would be represented by two V machine counters: C_1 and C_2.
The values of the two counters would satisfy

$$C_1 + C_i = T.$$

Here is how the operations were performed:

1. To add one to the counter C, one would add 1 to C_1 and subtract 1 from the
   counter C_2;
2. To subtract one from the counter C, one would subtract 1 from C_1 and add 1 to
   the counter C_2;

Note, this always preserved the invariant that $C_1 + C_2 = T$.

The key issue left is how to test whether or not the counter C is 0? Well it is 0
if and only if $C_1 = 0$ and $C_2 = T$. Thus, to tell that the counter C was 0 we would

attempt to subtract T from the counter C_2. Note, this would work only if the counter C is 0.

There is one more idea: how do we subtract T from a counter, especially since T could be a huge number. We do this by using a smaller counter that can count to $T^* = \sqrt{T}$ and a double loop that looks like this:

for $i = 1, \ldots, T^*$ do
for $j = 1, \ldots, T^*$ do
subtract 1 from C_2 and add 1 to C_1;

This loop can be implemented by a clever use of the previous counters, and relies on the fact that there is a reliable 0 test for smaller counters.

21.5 Abelian Semi-groups

As soon as Albert understood my proof he asked whether my V machines were reversible. I thought for a while, and we both worked out that they were. This meant that we could prove the following theorem:

Theorem 21.3. *The word problem for commutative semigroups is* EXPSPACE *hard.*

The idea was to suppose that the semigroup had three generators. Then an element of the semigroup is described by

$$a^x b^y c^z$$

where x, y, z are natural numbers. A semigroup is defined by equations on the generators such as
$$abc = a^3 c^2.$$

These equations are reversible in the sense that both the following transformations are allowed:
$$abc \rightarrow a^3 c^2 \text{ and } a^3 c^2 \rightarrow abc.$$

Since the V machines were reversible, the word problem for abelian semigroups could be shown to have the same EXPSPACE lower bound.

The details of this were worked out by Meyer and his graduate student E. Cardoza [28]. They published another paper without me; we had planned a joint paper, but I had a serious illness in my family and could not help with the paper.

21.6 Upper Bounds

Recall that after the lower bound proofs the following was proved:

Theorem 21.4. *The reachability problem for Vector Addition Systems is decidable.*

This theorem has an interesting story of its own.

The history of the proof falls into three acts. In act I, as with many hard problems, the theorem was claimed a number of times: each of these claims were found to be incorrect. For example, George Sacerdote and Richard Tenney–two theorists– claimed a proof that eventually was shown to have lemmas that failed even in the one-dimensional case. The reachability problem is a tricky problem, and intuition about it was often faulty. I am terrible at geometric reasoning myself, and I can understand how people could easily be mistaken.

In act II two proofs were found: the first by Ernst Mayr, and shortly thereafter a proof by Rao Kosaraju [83, 104]. Many did not believe Mayr's proof–it was un-clear in many places. All of these proofs are long complex arguments, with many interlocking definitions and lemmas, and understanding them is difficult. Kosaraju created his proof because he felt Mayr's proof was wrong. You can imagine there was an immediate controversy. Which proof was right? Both? Neither? There were claims that Kosaraju's proof was just Mayr's proof, but written in a clear way. The controversy got messy very quickly.

Finally, in act III there is now a new proof that seems much cleaner, and appears to be correct due to Jérôme Leroux [89].

21.7 Open Problems

The main open problem is to improve the lower bound on the reachability problem for VAS's. There is currently a huge gap between the lower bound of EXPSPACE and the upper bound, which is not even primitive recursive. The current construction of the lower bound is tight, and I see no way to improve it. But, perhaps you will see something that we all have missed.

21.8 Notes

This appeared as an-expspace-lower-bound in the blog. Luca Aceto who related a great dinner discussion he had with Meyer. He also remarked that Meyer is still a theorist, whether he works on complexity or on semantics. Jérôme Leroux and Joel Ouaknine agreed with Luca about who is a theorist.

Chapter 22
Randomness has Unbounded Power

Anne Condon is a theoretician who did fundamental work early in her career in complexity theory. She then moved into the area of DNA computing, and next to the more general area of algorithms for biology. Her work in all these areas is terrific, and she is a wonderful person to work with on any project. I had the privilege of working with her on a number of papers. I will talk about one of these.

The reason I want to present our result is twofold. First, it shows that randomness can convert a computational model that is very weak, into one that has the power of the halting problem. If you do not know this result you might expect that randomness can increase the power of a computational model, but that it can cause this huge a jump in power is a bit surprising. The second reason I want to present this result is it uses some interesting tricks that might be useful elsewhere in complexity theory. We use an old but important result of Minsky–yes the Minsky of AI fame, a clever theorem of Frievalds, and the machinery of prover-verifiers.

22.1 The Problem

I still remember visiting Madison when she was a member of the Computer Science Department. We chatted about her work, and she asked me a simple question about Markov Processes. The question was this: Suppose that we have two $n \times n$ stochastic matrices, call them A and B. Also suppose that v is a fixed n dimension probability vector. How hard is it determine whether or not there is a sequence of matrices $M_1, M_2, \ldots, M_t$, where each $M_i \in \{A, B\}$, so that the first coordinate of

$$M_1 M_1 \cdots M_t v$$

is above $2/3$? Or so that the first coordinate is always below $1/3$?

I was immediately intrigued and began to work on the problem. At first I guessed that the problem must be decidable. I had previously worked on problems that were close to this and shown that they were decidable. But I quickly began to hit the wall, nothing seemed to work. Then, I read Frievalds' beautiful paper on probabilistic finite state automata [52]. I quickly saw that it might be possible to improve his result to eliminate the need for two-way tapes.

22.2 Probabilistic Finite State Automata

A *1-way probabilistic finite automaton* (1pfsa) is just a finite state automata that has the ability to flip a fair coin. The machine as usual reads the input character, and then based on its state and coin flip enters a some (other) state and moves the input head to the right. The machine has accept and reject states. The probability that an input tape is accepted (rejected) is defined in the obvious manner. Such machines have been studied since the 1960's. Some of earliest results are due to Michael Rabin and to Azaria Paz. See Anne's survey for more details [36]. A *2-way probabilistic finite automaton* (2pfsa) is just a probabilistic finite state automaton that has instead a 2-way read-only input tape. The ends of the input tape are marked as usual.

The connection with Anne's problem should be clear. If the alphabet is binary, then the matrices A and B encode the operation of the automata in the obvious manner. So her question was really: Given a 1pfsa could we decide whether or not there was some input tape so the machine accepted with probability $2/3$, or did all inputs get rejected with probability $1/3$?

Note, this was a kind of promise problem: the automata was assumed to operate so that for any input either the accept probability was bigger than $2/3$ or less than $1/3$. This is, of course, a natural assumption, think of it like the assumption that defined BPP. All our probabilistic automata, 1-way or 2-way, will have error bounded away from $1/2$.

Rusins Freivalds (1981) had studied 2pfsa and proved the following surprising theorem [52]:

Theorem 22.1. *The language $\{a^n b^n \mid n \geq 0\}$ is accepted by a 2pfsa with bounded error.*

The proof idea is central to our extension so we will prove it in detail. Suppose that the input tape contains $a^n b^m$. We can assume this since this property is easily done by a finite state automata without any coins. The automaton will then use fixed size counters and check that $n \equiv m$ mod 1000. This also can be done by a finite state automaton. If the input tape fails either of these test, then we reject.

So we can assume that the input is of the form $a^n b^m$ with $n \equiv m$ mod 1000. Freivalds' clever idea is now to make repeated passes over the input tape. During each pass he will flip a fair coin for each a. If *all* of these come up heads, let's call it a *win for a*. During the same pass he with also flip a fair coin for each b. If *all* of these come up heads, let's call it a *win for b*. There are four possibilities: if neither a nor b get a win, then do nothing. If both get a win also do nothing. Suppose that a gets a win and b does not. Then, increment a special counter C_a: it starts out at 0.

In a similar manner, if b gets a win and a does not, then increment another special counter C_b; it also starts at 0.

When one of these counters gets to 100 stop. Suppose it is the C_a counter. Then, check the value of the counter for C_b. If $C_b > 0$ declare that $n = m$; otherwise, declare that $n \neq m$. That's the whole algorithm.

The question is why does this work? Suppose first that $n = m$. Then, clearly the chance of an a win is exactly the same as an b win. So the chance that $C_a = 100$ and C_b is still 0 is small. Suppose next that $n > m$. The same argument will work if $n < m$. We claim that the probability we make an error here is small: the only way we can make an error is to get $C_b = 100$ while $C_a > 0$. But the key point is that since $n > m$ and $n \equiv m$ mod 1000, $n \geq m + 1000$. So the probability of an a win is 2^{1000} less likely than a b win *independent* of n. This shows that the algorithm works. Pretty cool idea, I think.

22.3 Freivalds Main Theorem

Freivalds proved another theorem:

Theorem 22.2. *Every r.e. language is accepted by a 2pfsa with bounded error.*

Freivalds used that the power to count is enough to simulate a counter machine with two counters. Since these machines are universal by a theorem of Minsky [107] he is done. The details of going from counting, i.e. accepting the language $\{a^n b^n \mid n \geq 0\}$, to this theorem are a bit technical, but not very hard. I will discuss counters machines in more detail in a moment.

22.4 Sketch of Our Proof

Condon and I proved this [37]:

Theorem 22.3. *Every r.e. language is accepted by a 1pfsa with bounded error.*

The big difference is now the automaton is 1-way. This seems at the face of it a major problem. Freivalds' construction depends strongly on repeatedly going over and over the *same* input. How can we simulate his method if we get only one pass over the input?

The answer is we view the problem as a prover-verifier problem. Suppose a prover has a series of string of the form

$$a^{n_k} b^{m_k}$$

for $k = 1, \ldots$. The prover is consistent in the following sense: either all $n_k = m_k$, or none of the $n_k = m_k$. Our job is to use only a 1pfsa to tell which situation we are in: the equal case or the unequal case. Note, the prover can still be malicious:

1. The values of the n_k and m_k can vary as long as he is consistent;
2. In the unequal case the prover can make $n_k > m_k$ or $n_k < m_k$: there is *no* requirement to be consistent from one index to another.

This last ability destroys Freivalds' method. He needed to have a kind of race between a and b. Now that is thrown out, since the prover can vary which is larger.

The new trick is to play a more complex game with the prover. We think of the proving as presenting us with $a^n b^m$ each time. As before we can assume the string is always of this form and $n \equiv m \bmod 1000$. We know either $n = m$ always or $n \neq m$ always.

The idea is to find a "game" like Freivalds', but it must be symmetric: so no matter whether the prover in the unequal case makes $n > m$ or $n < m$ we will discover it. The basic insight is to flip fair coins again but do three types of games:

1. one game wins if $n + m$ coins come up heads;
2. one game wins if $2n$ come up heads;
3. one game wins if $2m$ come up heads.

In the equal case all have the same probability. In the unequal case, the one for $n + m$ is much more probable than one of the others. We can exploit this to get the theorem.

22.5 Minsky's Theorem

Marvin Minsky is a Turing Award winner and one of the founders of AI. But he also–early in his career–proved some very basic and pretty theorems. Some are so old that I am afraid many of you may not even know them. He proved the following beautiful theorem.

Theorem 22.4. *Every r.e. language is accepted by a two counter machine.*

A two counter machine has two unbounded counters and a deterministic finite state control. The machine also has an one way input tape. The machine can test a counter to see if it is zero, and add or subtract 1 from a counter. Subtraction is allowed only when the counter is positive. Minsky's theorem is that such a two counter machine is universal [107]. The key is he can show how two counters can simulate any fixed number of counters, which in turn can simulate a Turing Machine. The proof is quite clever and if you have not seen it before I think you may find the proof quite neat. The idea is to use

$$2^a 3^b 5^c 7^d$$

to represent the state of a four counter machine. You keep this value in one counter, the other is used just for operations. Suppose, we want to add one c: we just decrement the first counter by 1 and add 5 to the second counter. When the first is 0 the second will contain the value

$$2^a 3^b 5^{c+1} 7^d.$$

Testing for zero in a counter is the same trick; subtracting is also similar.

22.6 Open Problems

The method used here to count is clearly not very practical. However, there is something basic about the tricks used, I and believe they could play a role elsewhere. Also are there additional restrictions that we can place on the transition matrices A and B so that the machine has a decidable problem? What if they commute? Or is there some other property that will allow us to decide whether they accept a string or not?

22.7 Notes

This appeared as the-unbounded-power-of-randomness in the blog.

Chapter 23
Counting Cycles and Logspace

Ken Regan is a great theorist, and probably the best chess player in the world who can correctly state what P=NP means. Certainly, the best who can give the proof Toda's Theorem.

I want to talk about a connection between graphs, cycles, and permutations.

Ken is a very strong chess player. Very. He is an International Master (IM), which is the rank just below Grandmaster (GM). His international rating is 2405—ratings in the 2400's are typical for IM's, over 2500 for GM's. I am impressed, when I played in high school my rating started with a 1 and the next digit was not large.

Ken has recently been working on some fascinating projects on detecting cheating in chess tournaments. The issue is simple: is a player using a computer program to help make the moves? As you probably know computer programs are terrific today. Top level programs are matched very well against the best chess players in the world. They are especially strong in tactical positions that routinely arise in real matches. So access to a computer would yield a great advantage to the player.

During the 2006 Kramnik-Topalov world championship match cheating accusations against Kramnik were made. Ken was asked to try to decide how one might evaluate such accusations—that a player is making "too many" moves favored by a computer program, in this case the Fritz 9 program. Within days he posted his conclusion that Kramnik's high coincidence to Fritz and another program, Rybka, was due to the unusually forcing nature of the games, Game 2 in particular. As Ken says,

> If you force play and limit your opponent's options, don't be surprised if he matches a computer's moves!

Ken deserves a thanks from all chess fans, at all levels, for his quick and insightful analysis of this controversy. Now let's now turn to the result that I want to highlight.

23.1 Number of Cycles of a Permutation

The result concerns permutations. Of course a permutation is nothing more than a mapping π from a finite set S to itself that is one-to-one. The set is usually taken to be $1, 2, \ldots, n$ for some n, but we will allow any set S.

The problem that I want to discuss is: does π have an even or an odd number of cycles? Every permutation can be uniquely written, up to order, as a collection of cycles. The number of cycles of a permutation π will be denoted by $\mathsf{cycle}(\pi)$.

Computing the number of cycles of a permutation seems to a *sequential* property. I am not exactly sure how to state this precisely, but the obvious algorithm requires that you pick a point $x \in S$, and then trace out its cycle:

$$x, \pi(x), \pi(\pi(x)), \ldots, x$$

Then, move on to the next cycle, and eventually discover the cyclic structure of π. This yields, of course, the exact number of cycles of the permutation.

There is another way to compute the number of cycles of a permutation modulo 2 that is more indirect. This is the trick that I want to highlight:

Theorem 23.1. *For any permutation π,*

$$\mathsf{cycle}(\pi) \equiv |S| + \mathsf{inv}(\pi) \bmod 2$$

Here $\mathsf{inv}(\pi)$ is the number of inversions of the permutation π: an inversion is a pair of x and y in S so that $x < y$ and yet $\pi(x) > \pi(y)$. Here is an example: Consider the permutation that maps,

$$1 \to 2 \text{ and } 2 \to 3 \text{ and } 3 \to 1.$$

The number of cycles is 1, the cardinality of the set is 3, and there are 2 inversions. The inversions are $1, 3$ and $2, 3$. The theorem works, since $1 \equiv 3 + 2 \bmod 2$.

If we think of the cardinality of S as known, then the theorem really says:

$$\ulcorner \mathsf{inv}(\pi) \bmod 2 \urcorner \text{ determines the value of } \ulcorner \mathsf{cycle}(\pi) \bmod 2 \urcorner.$$

The reason I am excited by the theorem is that computing $\mathsf{inv}(\pi) \bmod 2$ seems, to me, to be a much less sequential task, than the usual method of determining the cyclic structure of a permutation:

• The usual method is sequential. It follows a point to another point and eventually determines its cyclic structure. Of course this method determines more than just the number of cycles modulo 2: it gets the exact number of cycles. This method clearly is a sequential method, which does not seem to be parallelizable in any obvious manner.

• The inversion method is easily made parallel. Have each "processor" take a pair of x, y from S, and then check whether they form an inversion: that is $x < y$ and $\pi(x) > \pi(y)$. This all can be done in one parallel step. Then, the processors each report back to a central processor, who computes the total number of inversions modulo 2.

23.2 Can We Do Better?

An obvious idea would be to try use similar methods to count the number of cycles modulo other numbers. In particular, can we determine the number of cycles modulo 3 with similar techniques? The answer Ken and I found is no. Roughly, a method that could do this would violate known complexity results. I will sketch the reason in the next sections.

23.3 Cycle Languages

I need to define a number of languages that will be important for our discussion. All the languages consist of directed graphs with some additional property. The input will always be of the form

$$(x_1, y_1), \ldots, (x_m, y_m),$$

where each x_i and y_i is in $\{1, \ldots, n\}$ encoded as binary strings. The meaning of (x, y) is that there is a directed edge from x to y.

The first is the language L_{cycle}, which consists of all inputs that encode a directed graph that consists of only vertex disjoint cycles. The next languages form a family of languages L_k and consist of all inputs that contain exactly k number of vertex disjoint directed cycles. Note,

$$L_k \subsetneq L_{cycle}$$

for any k. Also

$$L_{cycle} = L_1 \cup \cdots \cup L_k \cup \ldots$$

Finally, define L_{even} to be

$$L_2 \cup L_4 \cup \ldots$$

23.4 Basic Results

Some definitions are in order. The class AC^0 is the uniform version of the famous class of constant depth and polynomial size circuits with AND/OR/NOT gates, and the AND/OR gates have unbounded fan-in. Also $ACC^0[m]$ is the uniform class that also allows gates that compute mod m. The class L denotes logspace. The following two theorems are straightforward.

Theorem 23.2. *The language L_{cycle} is in* AC^0.

Proof. The key is that a directed graph has this form if and only if the in-degree and out-degree of each vertex is 1. Clearly, this can be checked as required. For example, the vertex v has in-degree at least 1 is equivalent to there is *some i* so that (x_i, y_i) is an edge and $v = y_i$.

Theorem 23.3. *For any $k \geq 1$, the language L_k is in* L.

23.5 Logspace Complete Problems

A language is *logspace complete*, for us, always with respect to AC^0 reductions. The following theorem and proof is based on Stephen Cook's and Pierre McKenzie's classic paper—"Problems complete for deterministic logarithmic space" [41].

Theorem 23.4. *It is* L*-complete to tell L_1 from L_3, or any L_k from L_{k+2}. Hence none of these tasks is in* $ACC^0[p]$, *for any prime p.*

The key parts of the proof are already in Cook-McKenzie; Ken and Pierre himself extracted them while the former was visiting the latter in Montreal. The case $k = 2$ reverses the intuition I'd had that the fact of telling the number of cycles modulo 2 meant it ought to be possible modulo 4, or modulo 8.... I had tried a wide range of graph and permutation tricks, all based on the fact that constant-depth networks of mod-2 gates can compute values mod-4. With hindsight this seems silly, and I could have had a correct intuition with better knowledge of Cook-McKenzie. The point is that research is often a game of competing intuition, and keeping one's mind open is often as important as trying to make something work.

23.6 Even Cycle Theorem

Another way to state Theorem 23.1 is the following.

Theorem 23.5. *The language L_{even} is in* $ACC^0[2]$.

A corollary of this is that L_{even} cannot be L-complete. This is an unconditional result, because otherwise it would violate known lower bounds on $ACC^0[2]$. The bounds are the brilliant work of Roman Smolensky, who showed that $ACC^0[p]$ cannot contain $ACC^0[q]$ when p is prime and q is divisible by some prime other than p [123]. If L_{even} were L-complete, then $ACC^0[2]$ could count modulo 3, in violation of Smolensky's theorem. The same violation would happen if $ACC^0[2]$ could count cycles modulo 4. Simply put, $ACC^0[2]$ does not equal L—this isn't a conventional belief like P$\neq$NP, it's a fact.

23.7 Open Problems

Can the Even Cycle Theorem be used to solve some interesting open problem? It seems like a neat trick that ought to have some other uses. It says that the parity *character* of finite symmetric groups is always $ACC^0[2]$-computable—what can we say about the complexity of other characters?

23.8 Notes

This appeared as graphs-permutations-characters-and-logspace in the blog.

Chapter 24
Ron Graham Gives a Talk

Ron Graham is part mathematician, part computer scientist, part juggler, part magician–he is one of the most talented people I know. He has done terrific work in the theory of computation, and in this chapter I will talk about just one of his many great results.

Speaking of talks, I would like to share a story about "talks" that involve Ron. It concerns the work that I will discuss here, his proof that the Euclidean Traveling Salesman Problem is NP-hard. He presented this work at the ACM STOC Conference back in 1976. The work was joint with Mike Garey and David Johnson, who are the authors of the famous book "Computers and Intractability: A Guide to the Theory of NP-Completeness" [54].

In 1976 Graham used an overhead projector to present his talk: this was way before Powerpoint. Then, talks were prepared by writing on clear plastic transparencies called slides. The slides were placed on top of a flat glass surface of the projector, under which was a very bright lamp. The light passed through the slide and reflected off a mirror and then onto the screen. In this way the speaker could see the slide and the audience could see the slide's image projected on the screen.

Sometimes people would point to places on the slide with a pen or even with their fingers in order to highlight some key part of the slide. This would, of course, appear on the screen as a dark shadow, since they pen or figure were opaque. Pointing in this manner could be annoying if done poorly, but it was a common practice. Today people prefer laser pointers, but in 1976 just as there was no Powerpoint, there were no laser pointers. I imagine then a laser would have cost thousands of dollars and been the size of a small piano.

During Ron's talk he had many slides that had points and lines drawn on them. He used these to explain the complex series of ideas that were used to prove that the Euclidean Traveling Salesman Problem could encode all of NP. Often in these types of reductions a "gadget" was needed. This result was no different, and Ron used a small piece of cardboard that he could move around the slides to highlight important parts of the construction. One advantage back then was that speakers could use two projectors, so that each projector could have a different slide on it. This was extremely useful as the speaker could leave a complex slide up while presenting the next one.

Ron was doing exactly this. He had two projectors going and was well into his talk when a miracle happened. He was moving the cardboard piece around one slide—you could see on the screen not only the piece but his fingers that were moving it. He then calmly walked over to the second projector to present another slide. *But his fingers were clearly still on the first projector.* I remember vividly how it slowly dawned on the room that the fingers on the first projector were not Ron's. They were fake. It was a great trick and caused an uproar of laughter. What a cool stunt.

24.1 A Problem with The Traveling Salesman Problem

Graham's paper [53] proved that the Euclidean Traveling Salesman Problem (TSP) was NP-hard. But it left open whether the problem is NP-complete. The difficulty hinges on a beautiful problem that is still open over thirty years later. The problem is this: Can we tell whether or not the inequality

$$\sqrt{a_1} + \sqrt{a_2} + \sqrt{a_3} + \cdots + \sqrt{a_n} \le k$$

is true in polynomial time? Here $a_1, a_2, \ldots$ and k are integers. If this could be determined in polynomial time then the Euclidean TSP would be in NP and so it would be NP-complete. Recently, Michel Goemans posted this problem on the website called the "problem garden." This is a site that is trying to compile a list of open problems for mathematics and computer theory. Take a look at it. Even better post your open problems there. Goemans also pointed out that this problem is of importance to our understanding of semi-definite programming.

24.2 Partial Results

There has been some partial results on the sum of square roots problem (SSP). First, there are results from computational geometry that has shed some light on related questions. A typical result is to look at how close the following can be to zero:

$$|\sqrt{a_1} + \cdots + \sqrt{a_n} - \sqrt{b_1} + \cdots + \sqrt{b_m}|.$$

Second, there are some very pretty results by Eric Allender and his co-authors on a complexity approach to the SSP. They show that the problem is in the so-called counting hierarchy. I will not go into details now, but for details look at their paper [3].

24.3 An Exact Sum of Squares Problem

I would like to present a result that uses a very important technique due to Zhi-Zhong Chen and Ming-Yang Kao. They used the method in a great paper "Reducing randomness via irrational numbers" [31].

I will present their method as a tool for solving the exact sum of square problem (ESSP): this is the problem of whether or not

$$\sum_k a_i \sqrt{b_i} = 0$$

where a_i are positive or negative integers and b_i are only positive integers. The problem is more general than SSP, since we allow coefficients on the square roots; it is simpler since we only ask that you determine whether or not the sum is *exactly* equal to 0. The technology that Chen and Kao have can only work for equalities; it appears not to be of any use for inequalities. I say *appears*–perhaps there is a clever way to extend their method to inequalities. I do not see how, but perhaps you will.

Their method uses some basic facts from field theory and the theory of algebraic numbers. Perhaps the lesson for us is that often classic mathematics can make important contributions to the design of algorithms.

I think the best way to present their idea is through a simple example: suppose that

$$\alpha = 112\sqrt{3} + 134\sqrt{5} - 201\sqrt{6}.$$

This is a toy example. If you calculate the value of α you will get about 1.27. Not too hard to see that its non-zero. But consider the following number

$$\beta = 112\sqrt{3} - 134\sqrt{5} + 201\sqrt{6}.$$

The value of beta is about -597. Clearly, it would be easier to calculate β to low precision and see that it is non-zero, then to do the same for α.

The first observation is that $\alpha = 0$ if and only if $\beta = 0$. The reason is that β is a *conjugate* of α. Algebraic numbers have conjugates, and one conjugate is zero if and only if all the conjugates are zero. Moreover, in the case of numbers that are made up of only square roots, conjugates are easy to determine: in general the numbers $\sqrt{a}$ and $-\sqrt{a}$ are conjugates.

Thus, if we are testing for zero we can use any conjugate that we wish; one is just as good as another. The second observation is that often conjugates can have wildly different absolute values. We just gave an example where one was over 100 times larger than the other. Chen and Kao can prove that this is the typical situation. For a more precise statement look at their paper, but the key insight is that of all the conjugates most will be relatively large. So the way to solve the ESSP is this: pick a *random* conjugate of the sum of squares roots that you wish to test, then compute it to a polynomial number of bits of accuracy. If the answer is small, then say the sum is exactly 0; otherwise, say its non-zero. The probability that this works is quite high. Very neat idea.

For those who want a bit more details here goes. The number α not only is an algebraic number but is an algebraic integer. The norm of α, usually denoted by $N(\alpha)$ must be at least 1 unless α is zero. But the norm is just the product of all the conjugates of α. An easy argument then shows that since the absolute values of all the conjugates are not too big, there cannot be too many small conjugates.

24.4 Open Problems

The main open problem is to show that the SSP can be done in polynomial time. However, this is over-kill: you actually only need to show that the SSP can be done in nondeterministic polynomial time. This would be enough to show that TSP is NP-complete. This is a simple observation, yet I do not know of any results that have tried to exploit this insight. Perhaps, you can use it to solve Graham's original problem.

A second even smaller observation is this: consider an SSP,

$$\sqrt{a_1} + \sqrt{a_2} + \sqrt{a_3} + \cdots + \sqrt{a_n} \le k$$

that arises from a TSP problem. It this case the square roots arise from the calculation of Euclidean distances. If one point is (r,s) and another is (u,v), then their distance is

$$\sqrt{(r-u)^2 + (s-v)^2}.$$

After clearing denominators the values of a_i will be special integers. In particular, they will always be the sum of two squares. Fermat essentially proved that x is represented as the sum of two squares if and only if in the prime factorization of x, every prime of the form $4t+3$ occurs an even number of times. The final thought, is can we exploit this fact? Can the special nature of the square roots be useful?

24.5 Notes

This appeared as ron-graham-gives-a-talk in the blog. Timothy Chow pointed out I incorrectly cited Goemans as placing the problem in the Problem Garden. Also Omar Camarena asked for more details on how random conjugates are selected for numbers with unknown factorization. The answer is to use greatest common divisors, but I thank him for this nice question.

Chapter 25
An Approximate Counting Method

Larry Stockmeyer was one of the great theorists, who worked on areas as diverse as lower bounds on logical theories, computational complexity, distributed computing, algorithms, and many other areas. He passed away in 2004, and is terribly missed by all who knew him.

Let's talk about one of Larry's beautiful results: an approximate counting method. I have always loved this result.

Larry was quiet, but beneath his quiet exterior he had a tremendous sense of humor. He was also one of the most powerful problem solvers that I ever had the pleasure to work with. Larry was also extremely accurate—if Larry said that X was true, you could bet the house that X was true. He was amazing at finding loose ends in an argument, and usually then finding a way to tie them up.

I once proved a result on the complexity of univariate polynomials, and I showed a draft of the paper to Larry [93]. A few hours later he came into my office—I was visiting IBM Yorktown at the time—with my paper all marked up. Larry then started to give me very detailed comments about the substance of the paper. After a while I started to get nervous; I was starting to wonder whether or not the main result of my paper was correct or not. I finally asked Larry directly: "is my theorem correct?" He said,

> oh yes, but here you need to argue that ...

I was quite relieved. Larry and I went on to work together on other results on the complexity of polynomials.

A year later, again at IBM Yorktown, we had two visitors George Sacerdote and Richard Tenney, who came to IBM and gave a very technical talk on their solution to the *Vector Addition System* (VAS) reachability problem—see Chp. 21. It was a long, almost two hour talk, that was attended by most of the IBM theory group. The reachability problem had been claimed before, and I had worked hard on the problem in the past—with no success. So I was very interested to see if they had really solved the problem.

When the talk was over, we all started to leave the conference room and head to the IBM cafeteria for lunch. I stood right at the door and did an "exit poll"—as each theorist left the room I asked them what they thought about the proof. Did they believe it? Did it look like it might work? Most said that it sounded plausible. A few were even more positive: one member said that he was convinced that the proof would work. Personally, I thought that I had not heard anything new, but neither had I heard anything wrong. I was on the fence.

I then asked the last to exit the room, Larry, what he thought about the proof. Larry said,

> I did not understand a word that they said.

Larry was right on the money. The proof was wrong, but it took a lot of work, by many people, to eventually find the holes in their attempted proof.

Let's turn to Larry's result on approximate counting.

25.1 Approximate Counting

Suppose that $C(x)$ is a circuit of size polynomial in n with inputs $x = x_1, \ldots, x_n$. Then, a natural question is: How many x satisfy $C(x) = 1$? This is well known to be a #P complete problem, so computing the exact answer is certainly at least as hard as NP. What Larry looked is how hard is it to approximately determine the number of x so that $C(x) = 1$? Let $S = \{x \mid C(x) = 1\}$.

A key observation is that there is an "amplifier" lurking here: if we can in general determine $|S|$ to within a factor of 2 in polynomial time, then we can determine it to within factor $1 + \frac{1}{n^c}$ for any constant c also in polynomial time. This can be proved by a simple amplification argument. The idea is this: create a new circuit $D(x, y) = C(x) \wedge C(y)$ where x and y are disjoint inputs.

Then, $D(x, y)$ has $|S|^2$ inputs that it accepts. If you know $|S|^2$ to a factor of 2, then you know $|S|$ to a factor of $\sqrt{2}$. An m-fold version of this will yield a approximation of

$$2^{\frac{1}{m}} \approx 1 + O(\frac{1}{m}).$$

This is a perfect example of the power of even a simple repetition amplifier—see Chp. 40.

Thus, the key problem is to determine $|S|$ to within a factor of 2. Larry proved [125],

Theorem 25.1. *There is a randomized algorithm that determines $|S|$ to within a factor of 2 in polynomial time, provided it has access to an NP-oracle.*

Note, there is no way to remove the need for the oracle, without a breakthrough, since it is easy to construct an NP-hard problem that either has no solutions or many solutions. Thus, determining the answer to within a factor of 2 without the oracle would imply that P=NP.

25.2 Sketch of Larry's proof

Larry's proof uses some ideas that had been used earlier by Mike Sipser, but he added several additional insights. See his paper for the details, but I will give a quick overview of how the counting method works.

Let $C(x)$ be the circuit and let S be the set of inputs x so that $C(x) = 1$. Suppose that our goal is really modest: we want to know if $|S|$ is really large, or really small. Take a random hash function $h : \{0,1\}^n \to \{0,1\}^m$ where m is much smaller than n. Then, check to see if there are two distinct x and y so that they form a "collision",

$$h(x) = h(y), \tag{25.1}$$

and both are in S. If $|S|$ is small then, it is likely that (25.1) will have no solutions, but if $|S|$ is large, then it is likely to have many solutions. The key to make this work is careful analysis of the probabilities of collisions for the given hash functions. This can be worked out to prove the theorem. Note, the detection of a collision requires an NP oracle call:

are there two distinct x and y such that $h(x) = h(y) = 1$ and both are in S?

25.3 Counting and Finite Automata

I love finite state automata (FSA)—as you probably know. The following theorem is not as well known as it should be, in my opinion at least:

Theorem 25.2. *Suppose that M is a m-state deterministic finite state automaton. Then the number of length n inputs that are accepted by M can be computed exactly in polynomial time in n and m.*

Thus, for FSA, the problem of counting the number of accepting computations is easy.

A proof sketch is the following. Let M' be a new automaton that simulates M on all inputs of length n, and has the property that the state diagram is acyclic. Essentially, M' replaces each state s of M by the state (s,i) where $i = 0,\dots,n$. For example, if $a \to b$ was a transition on input 0, then

$$(a,i) \to (b,i+1)$$

is a transition for input 0 for all i. We have simply unrolled the automaton M to avoid any loops: this cannot be done in general, but is fine if we are only concerned with inputs of a fixed length n.

The algorithm then inductively labels each state (a,i) with the number of length-i inputs that reach this state. To label a state $(b,i+1)$, we take each state (a,i) with arcs to $(b,i+1)$, and add the number of (a,i) *times the number of arcs from a to b*. The number of accepting computations of length n is then the sum of the numbers for (b,n) with b an accepting state of the original M.

25.4 Open Problems

Can we improve the result of Stockmeyer on approximate counting? For specific problems there are better results known of course, but I wonder, can we improve his counting argument? Of course if P=NP, then approximate counting would be in polynomial time, but can it be in NP?

   Another natural question that I think might be worth working on is this: pick a complexity class that could be weaker than polynomial time and see what the cost of approximate counting is for that class. I have given a simple example, above, where exact counting is easy. There is some quite nice work on approximate counting for certain simple classes of circuits.

25.5 Notes

This appeared as stockmeyers-approximate-counting-method in the blog. Ken Regan discussed the relationship of approximate counting to quantum computation. Ron Fagin had a wonderful comment:

> About 10 years ago, I found a false proof that P does not equal NP, and it took me a few days to find my bug (I was pretty excited for a while!). Even though I had already found my bug, I showed the false proof, just for fun, to various famous members of our theory community, to see if they could spot the bug (I did tell them that there existed a bug, and that their challenge was to find it). Only two of them were able to find the bug, and they both found it pretty quickly—Larry Stockmeyer and Moshe Vardi. Amusingly enough, the bug was, of all things, a misuse of Fagin's Theorem! Maybe the other people I showed the proof to just trusted that I surely wouldn't misuse my own theorem.

Chapter 26
Easy and Hard Sums

Jin-Yi Cai is one of the top researchers in complexity theory. Jin-Yi is one of the greatest pure problem solvers that I have had the pleasure to work with. He has solved a number of famous open problems, starting with his thesis work under the great Juris Hartmanis, continuing with his work on several long-standing complexity conjectures, and more recently extending the beautiful work of Leslie Valiant on *holographic* computation [129].

Let's talk about "doing sums." The problem is computing the sum of elements from an exponential size set in polynomial time. These problems arise in many parts of computer science, yet we still do not completely understand when they are easy to compute and when they are hard to compute.

Cai is a joy to work with and do research with. I am trying to think of a funny or strange story, but I cannot think of anything that unusual about him. There is one story of how we worked hard and solved a problem, and then got completely passed by a much more clever result of Richard Cleve [33].

Recall that Barrington's Theorem (see Chp. 19) shows that polynomial length bounded width computations can compute all of NC^1 [14]. David Barrington did not care about about the exact polynomial-length, he only cared that it was polynomial. Jin-Yi and I tried to improve his construction. The issue is this: suppose that F is a formula of size $|F|$ with all paths to leaves the same length. Barrington shows that this can be computed by a bounded width computation of length order $|F|^2$, when F is balanced.

Cai and I tried to reduce the exponent 2 on $|F|^2$. After a great deal of work, we were able to show how to improve Barrington's initial bound from 2 to 1.81. Our method used a complex "gadget" and some other ideas.

As soon as our result was published, others started to work on the problem. Cleve wiped us out. He proved that the exponent could be 1.7 or 1.6 or even better $1 + \varepsilon$ for any $\varepsilon > 0$. He saw much more deeply into the problem than we did; he set up a pretty framework that avoided any ad-hoc gadgets, which allowed him to essentially solve the problem completely.

26.1 Summation Problems

The general summation problem over a field F is: find the value of

$$S = \sum_{x \in A} f(x),$$

where A is an arbitrary finite subset of F^n and $f(x)$ is an arbitrary function from F^n to F. When the field is a subset of the complex numbers, it makes sense to talk about computing the value S with either additive or multiplicative error.

I find these problems interesting for several reasons:
• Summations arise naturally in many settings. However, the classic approach in mathematics is to get bounds on the value of summations, whereas computer scientists want algorithms that can compute them. That does not mean that general bounds are not related to computing their values, but general bounds are not enough.
• Summations arise as a model of quantum computation. If we could compute certain summations efficiently, then we could simulate quantum computations. This is likely too much to ask, but the question then becomes, when can a sum be computed fast?
• They yield some interesting mathematics, and I like the questions that arise. For a variety of reasons, summation type questions seem not to have received sufficient attention in complexity theory. Yet I would argue that they are a natural set of questions.

26.2 A Hard Family of Summations

Let me start by showing the following "folklore theorem." I have known it for a while, but cannot find a reference.

Notation: Hereafter, I and J will denote intervals of integers, i.e. I could be $[a,b]$, which is equal to

$$\{a, a+1, \ldots, b-1, b\}.$$

When I say $\sum_{x \in I}$, where I is an interval, I mean the summation to be over all integers x in the interval I. Also, the notation $|I|$ is the number of integers in the interval.

Theorem 26.1. *Let $r(x,y)$ be a rational function of degree at most 2. Computing the sum*

$$S = \sum_{x \in I} \sum_{y \in J} r(x,y)$$

is at least as hard as integer factoring.

Lemma 26.1. *Consider the following function $S(I,J,M,T)$ defined by*

$$S(I,J,M,T) = \sum_{x \in I} \sum_{y \in J} \frac{1}{T(xy - M) + 1}.$$

Assume $T > 3$. Then $S(I,J,M,T)$ is equal to $|A|$ where

$$A = \{(x,y) \mid x \in I \text{ and } y \in J \text{ and } xy = M\} + O(|I| \times |J|/T).$$

Proof. Define $r(x,y)$ to be

$$\frac{1}{T(xy - M) + 1}.$$

Then, if $xy = M$, then $r(x,y) = 1$. I next claim that if $xy \neq M$, then $|r(x,y)| \leq 2/T$. Suppose that $xy - M = k$. If $k > 0$, then the claim is easy to see. Thus, assume that $k < 0$ and that the inequality is false. Then,

$$|r(x,y)| = \frac{1}{|k| \times T - 1} > \frac{2}{T},$$

which implies that

$$T > 2|k|T - 2 \geq 2T - 2.$$

Recall that $T > 3$, hence this is a contradiction. This proves the claim.

It remains to calculate the contribution of each x, y to the sum $S(I,J,M,T)$. Each x, y with $xy = M$ contributes 1; each other x, y contributes at most $2/T$. The total number of x, y is clearly bounded by $|I| \times |J|$.

This proves the lemma.

Lemma 26.2. *Factoring of M can be reduced to a polynomial number of evaluations of the function $S(I,J,M,T)$.*

Proof. Let M be an n-bit number, and let $T = 2^{3n}$. Define the function $g(a,b) = S([a,b], [2, M-1], M, T)$. By the lemma, $g(a,b)$ is equal to the number of divisors $x|M$ so that $a \leq x \leq b$, plus an error term that is $o(1)$.

Start with $g(2, M-1)$. If this is $o(1)$, then M is prime. So assume that $g(2, M-1) \geq 1 + o(1)$. Then, by a binary search we can find a factor of M.

Two remarks about this theorem. First, we can weaken the assumption on computing $S = S(I,J,M,T)$ exactly. The argument actually shows that either $S \geq 1 + O(|I| \times |J|/T)$ or $S \leq O(|I| \times |J|/T)$. For T large enough, this shows that the method can handle either additive error of almost 1 or multiplicative error of exponential size.

Second, the theorem shows that the summation problem for very simple rational functions is hard. Using an unpublished–I believe–result of Joe Killian we can do much better. Killian proved the following:

Theorem 26.2. *Given an integer N, the following question is NP-hard: Is there a factor of N in the range $[a,b]$?*

The intuition for this result is that N could have many prime factors. Even if one is given the factorization of N, it could be hard to find a subset of the prime factors, whose product lies in the given interval. Another way to state this insight is that the problem is essentially a knapsack type question, where the values to be packed are the logarithms of the prime factors of N.

This result and the above methods yield the following stronger theorem:

Theorem 26.3. *Let $r(x,y)$ be a degree at most 2 rational function. Computing the sum S*

$$S = \sum_{x \in I} \sum_{y \in J} r(x,y)$$

is NP-hard.

This uses a proof similar to that of the previous lemma:

Proof. Let M be an n-bit number, and let $T = 2^{3n}$. Define the function $g(a,b) = S([a,b], [2, M-1], M, T)$. By the lemma, $g(a,b)$ is equal to the number of divisors $x|M$ so that $a \le x \le b$, plus an error term that is $o(1)$. Thus, determining if there is a factor in an interval reduces to one summation evaluation.

There is some interesting earlier work on related problems. For example, Leonid Gurvits and Warren Smith showed that computing the integral of rational functions that are given by straight-line arithmetic circuits is #P-hard [61]. The key to their result is the pretty formula for the permanent of an n by n matrix A,

$$\frac{1}{2\pi} \int_0^{2\pi} \prod_{j=1}^{n-1} h(p, \theta) d\theta,$$

where $p \ge 2$ is an integer and $h(p, \theta)$ is equal to

$$\sum_{k=1}^{n-1} a_{jk} \exp(ip^{k-1}\theta) + a_{jn} \exp(-i\frac{p^n - 1}{p-1}\theta).$$

26.3 An Easy Family of Summations

An interesting family of "easy" summations arises from a recent paper by to Jin-Yi Cai, Xi Chen, and Pinyan Lu (CCL) on graph homomorphisms: "Graph Homomorphisms with Complex Values: A Dichotomy Theorem" [137]. László Lovász first defined the notion of *graph homomorphism*, which generalizes many interesting graph properties such as: vertex cover, independent set, and graph colorings—to name a few. Thus, the paper of CCL, by allowing complex valued graph homomorphisms, covers many important graph properties. A natural question is, which properties are easy and which are hard?

They answer this question and prove a dichotomy-type theorem. Roughly speaking such theorems prove that all problems of a certain type can be divided into two classes: the first class are all in polynomial time, and the second are all NP-hard. The proof that the first class are in polynomial time relies on an interesting family of summations, which we will discuss in a moment.

You probably have seen dichotomy type theorems before. A very simple example is this: consider CNF SAT problems. Those with only two-clauses are always easy to solve. Those with three-clauses are, of course, NP-complete, in general.

The work of CCL is a much harder, a much deeper, and a much more powerful result. However, it is in the same spirit as the simple example I just gave. I will not be able to do justice to their beautiful paper, since it is a mere 111 pages—yes, that's not a typo, nor is it in binary. Read their paper for the details. Or at least browse through the paper–it contains many interesting ideas.

CCL must be able to classify certain summations that arise from finite rings, and decide which are easy to compute and which are not. The main result on the easy ones is this pretty theorem:

Theorem 26.4. *Let N be any positive integer, and let $\omega_N = e^{2\pi i/N}$. Let f be a quadratic polynomial in n variables $x_1, \ldots, x_n$ over $\mathbb{Z}_N$. Then the following sum*

$$Z_N(f) = \sum_{x_1, \ldots, x_n \in \mathbb{Z}_N} \omega_N^{f(x_1, \ldots, x_n)}$$

can be evaluated in polynomial time in n and $\log N$.

Note, when N is a prime, the result follows from known results. However, the full general case when N is an arbitrary number with unknown factorization is new.

These sums are quite interesting, and it is not obvious that they are easy to compute for several reasons. First, of course, the number N^n of terms in the sum is exponential. Second, raising a root of unity to the value of a quadratic polynomial is a term that varies in a complex manner. Finally, the problem is an exact evaluation: no error–the exact value is computed. The bottom line is that it is not at all clear why such a sum should be easy to compute.

26.4 Sketch of the Proof

The proof of the above theorem is fairly involved. The main idea is to show that it can be reduced to a class of sums that Gauss knew about. I will, with Cai's kind permission, include a sketch of the proof for the special case of N an odd prime power. This, I believe, gives the flavor of how such theorems are proved.

§

Define the *Gauss sum* as

$$G(a,N) = \sum_{t \in \mathbb{Z}_N} e^{\frac{2\pi i}{N} a t^2},$$

Gauss knew the value of these sums, and in closed form.

The main idea of the proof is to use induction on the number of variables. Let p be an odd prime, and $N = q = p^k$. We are trying to compute $Z_q(f)$. We assume

$f(x_1, x_2, \ldots, x_n)$ has the following form:

$$f(x_1, \ldots, x_n) = \sum_{i \leq j \in [n]} c_{i,j} x_i x_j + \sum_{i \in [n]} c_i x_i + c_0.$$

where all the $c_{i,j}$ and c_i are elements in $\mathbb{Z}_q$. We can assume the constant term in f is zero since it can be taken out of $Z_q(f)$ as a constant factor.

Note, if f was purely linear, then the sum could be easily written as a product of a simple sum and a sum over $n-1$ variables. Even though f is not linear, the strategy is to achieve this same goal. Thus, we can assume that f has non-linear terms.

For any non-zero $a \in \mathbb{Z}_q$, we can write it as $a = p^t a'$, where t is a unique non-negative integer, such that $p \nmid a'$. We call t the order of a (with respect to p). Since f has non-zero quadratic terms, let t_0 be the smallest order of all the non-zero quadratic coefficients $c_{i,j}$ of f. We consider the following two cases. There exists at least one square term with coefficient of order t_0, or not. Note, the fact this is over a ring, not a field, requires the complex considerations.

In the first case (I), without loss of generality, we assume $c_{1,1} = p^{t_0} c$ and $p \nmid c$ (so c is invertible in $\mathbb{Z}_q$). Then by the minimality of t_0, every non-zero coefficient of a quadratic term has a factor p^{t_0}. Now we factor out $c_{1,1}$ from every quadratic term involving x_1, namely from $x_1^2, x_1 x_2, \ldots, x_1 x_n$ We can write

$$f(x_1, x_2, \ldots, x_n) = c_{1,1}\left(x_1 + g(x_2, \ldots, x_n)\right)^2 + c_1 x_1 + \Delta,$$

where g is a linear form over $x_2, \ldots, x_n$ and Δ is a quadratic polynomial in $(x_2, \ldots, x_n)$.

By adding and then subtracting $c_1 g(x_2, \ldots, x_n)$, we get

$$f = c_{1,1}\left(x_1 + g(x_2, \ldots, x_n)\right)^2 + c_1\left(x_1 + g(x_2 \ldots, x_n)\right) + f'(x_2, \ldots, x_n),$$

where $f'(x_2, \ldots, x_n) \in \mathbb{Z}_q[x_2, \ldots, x_n]$ is a quadratic polynomial over $x_2, \ldots x_n$.

For any fixed $x_2, \ldots, x_n \in \mathbb{Z}_q$, when x_1 takes all the values in $\mathbb{Z}_q$, $x_1 + g(x_2, \ldots, x_n)$ also takes all the values in $\mathbb{Z}_q$. Thus,

$$\sum_{x_1, \ldots, x_n \in \mathbb{Z}_q} \omega_q^{f(x_1, \ldots, x_n)} = \left(\sum_{x \in \mathbb{Z}_q} \omega_q^{c_{1,1} x^2 + c_1 x}\right)\left(\sum_{x_2, \ldots, x_n \in \mathbb{Z}_q} \omega_q^{f'(x_2, \ldots, x_n)}\right)$$

$$= \sum_{x \in \mathbb{Z}_q} \omega_q^{c_{1,1} x^2 + c_1 x} \times Z_q(f').$$

The first term can be evaluated in polynomial time using Gaussian sums, and the second term is reduced to $Z_q(f')$ in which f' has at most $n-1$ variables.

Now we compute the sum $\sum_{x \in \mathbb{Z}_q} \omega_q^{c_{1,1} x^2 + c_1 x}$ using Gauss sums. If $c_1 = 0$, then this is already $G(c_{1,1}, q)$. Supposing $c_1 \neq 0$, we let $c_{1,1} = c p^a$ and $c_1 = d p^b$, where $p \nmid c, d$. Then we can show that if $a \leq b$, we can "complete" the square and the sum becomes (a constant multiple of) a Gauss sum. On the other hand if $a > b$, we can show that the sum is 0.

Finally, the second case (II) is reduced to the first case by a change of variable transformation. See their paper for all the details.

26.5 Open Problems

I believe that the following should be easy to compute, but I cannot prove it. Compute the sum

$$S = \sum_{x \in I} \frac{1}{rx + s}$$

within an additive error of 2^{-n}, where $I = [a, b]$ and a, b, r, s are at most n-bit numbers. Note, it is easy to compute the sum

$$\sum_{x \in I} p(x)$$

where $p(x)$ is a polynomial of degree n in time polynomial in n.

Another set of questions surround the work of CCL. Their algorithm works in polynomial time. Can we improve it to a "lower level complexity class?" What about, for instance, the class NC?

26.6 Notes

This appeared as computing-very-large-sums in the blog. Pascal Koiran and Qiaochu Yuan discussed whether or not certain sums should be easy to compute or not.

Chapter 27
How To Avoid O-Abuse

Bob Sedgewick is an expert on the analysis of algorithms–one of the best in the world. This is not too surprising since he is a Ph.D. student of Donald Knuth. Bob is also famous for his textbooks on algorithms, and more recently his monograph on the analysis of algorithms with Philippe Flajolet. This [48] is a required book for anyone trying to understand the performance of an algorithm.

I have many stories about Bob. One of the strangest is the time that we were offered a bribe. For the record we did not take the bribe–although the statute of limitations is probably long run out on us.

At the time Bob was the chair of the department of computer science at Princeton. For a variety of reasons there was a mathematician, we will call X, that was not going to be kept on at the Princeton mathematics department. This department is perhaps the best in the world, so that should not come as a shock.

The complication is that X was sort of a computer scientist. For reasons that I did not understand then, nor do I understand now, there were people in the political world that really wanted X to stay at Princeton. Powerful people. So far fine. It's nice to have friends in high places. The trouble began when Bob was "asked" to a meeting with them. I was brought along for support, and it turned out that was a good idea.

The meeting was strange. They spoke highly of X. We agreed and said that we had a mechanism for hiring faculty. For starters, X should send us a resume, which would start the hiring review process. They nodded, but asked why could we not just hire X. Bob repeated again and again that there was a process–and on it went.

Finally, one of the senior political types made it very clear that we personally would be very happy if we just hired X. No, he did not open a briefcase of cash, but his meaning was clear. Bob and I repeated the mantra, have X send in the resume. The meeting finally ended.

As soon as Bob and I were outside, we looked at each other and Bob said,

that was a bribe offer, no?

I agreed. A very strange affair all around. By the way X went elsewhere and did just fine.

27.1 How to Not to Abuse O-Notation

Bob has a simple idea that I find quite compelling. I will present it in his own words. His initial "this" refers to an algorithm that we were discussing via email.

This example brings to mind the idea of rampant misuse of O-notation, and an easy way to fix it that I've been teaching and using lately. Suppose that you think that the running time of one algorithm is $O(N \log N)$ and another is $O(N)$. With O-notation, you have nothing to test with an experiment. But in almost all cases where you are going to run an experiment, it is reasonable to use $\approx$ notation to express a hypothesis. In this case, you would hypothesize that the running time of the first is $\approx aN \log N$ and the second is $\approx bN$ for some constants a and b. So the ratio of the running times $R(N)$ should be $\approx (a/b) \log N$. Then if you run the same experiment for $10N$ and take the ratio $R(10N)/R(N)$, all the constants cancel and this ratio of ratios approaches 1. Under the hypothesis, running the experiment for larger N should give a better approximation to $R(N)$. Or, you can just run an experiment and solve for the constants to predict running time.

In short, experimental verification of hypotheses is easy with $\approx$ notation and impossible with O-notation. Despite that fact, O-notation is widely misused to predict and compare algorithm performance.

Bob then added, in his next email:

But actually log factors are not so important with this approach. For an unknown algorithm, we hypothesize $T(N) \approx aN^b$, find b by computing $T(2N)/T(N)$ for as large an N (and as many experiments) as we can afford, then use the actual running time to solve for a. This gives very accurate predictions. Essentially the log factor is absorbed into the constant, which is not so bad for large N.

27.2 Open Problems

Try Bob's idea out. Does this simple idea work on interesting problems?

27.3 Notes

This appeared as o-abuse in the blog.

Chapter 28
How Good is The Worst Case Model?

Claude Shannon is the father of information theory. While at Bell Telephone Labs—he wrote the seminal paper [122] that created the field in 1948. That is right, it was called " Bell Telephone Labs" in those days. His famous paper set out the basic framework of information theory that is still used today: a message, an encoder, a channel with noise, a decoder. Very impressive.

Shannon also proved the existence of "good" error correcting codes via a probabilistic existence argument. While we usually associate the probabilistic method with Paul Erdős, Shannon's use of a randomly selected code, was I believe, just ahead of Erdős. Of course, while random codes can correct errors at rates near the channel capacity, they are hard to encode and to decode. It is easy to get rid of the decoding problem as shown later, one can select a random linear code.

I want to discuss the Shannon model from a modern point of view, and also to relate it to one my pet open questions of theory: *what is the right complexity measure for algorithms?*

28.1 Worst Case Model Complexity

I love the worst case complexity model, and I hate the worst case complexity model. I love it because there is nothing better than an algorithm that works in the worst case. By definition, such an algorithm can handle anything that you can "throw" at it. All inputs are fine: clever inputs, silly inputs, or random ones. Worst case is great.

I hate the worst case model because it is too restrictive. Often we cannot find an algorithm that works on all inputs. Yet you may need an algorithm that works on the actual inputs that arise in practice. What do we do in this case?

There are several options, of course. We can throw up our hands and use heuristics, but then we cannot prove anything comforting about the behavior of the algorithm. Or we can use average case complexity. Whereby the inputs are assumed to come from a random distribution. Leonid Levin has worked on formalizing this model, and there are many interesting results [91]. However, I do not think that it is the right model either.

Inputs may not come from an all-powerful adversary, but are usually not randomly selected from some distribution either. There must be some model that is in between worst case and random. There are amortized complexity models, and even other models of complexity. Yet I feel that one of great open question in complexity theory still is:

How do we model inputs that better reflect what happens in the "real-world?"

While I wish I could tell you the answer, the best I can do is explain an idea that works in the area of information theory. So let's look more carefully at what Shannon did. Perhaps the idea that I have that works in information theory can be generalized to other areas of theory.

28.2 Shannon's Model

Recall Shannon's model. In modern terms, Alice wants to send Bob a message M via a channel. The channel may flip some of the bits of Alice's message, so Alice cannot simply send the message down the channel. Instead, she encodes the message into $E(M)$. Bob receives $E(M) \oplus R$ where R is the noise that the channel added to the message. Bob's job is to decode:

$$D(E(M) \oplus R) = M'.$$

Bob computes via his decoder a message M'. The claim is that provided the channel did not change too many bits of Alice's message, $M = M'$.

Shannon's existence proof is that there *exists* an encoder and a decoder that work, provided the number of bits changed is below a certain rate.

28.3 Genius Channels

There is a problem with the Shannon model, I think. He allows the channel to be worst case: it can change any bits that it likes. The decision which bits are changed can be based on any rule, i.e. there is no computational limit on the channel's ability. It is hard to believe that a channel is a "genius." Such a channel can read all the bits of the message, think for as long as needed, and then decide which bits to corrupt. Noise in many situations is driven by physical effects: sun-spots, electric effects, reflections of signals, and so on. Thus the idea is to replace the all powerful channel by a more restricted one.

28.4 Dumb Channels

In information theory often results are proved for so called memoryless channels: these channels flip a biased coin and if it is heads they change the bit, else they leave it uncorrupted. From the complexity point of view they are very weak; they have no memory and cannot do anything very powerful. Let's call them "dumb channels."

Not too surprisingly it is easier to design codes that work for dumb channels.

28.5 Reasonable Channels

My suggestion is to replace a channel by an adversary that has limited resources. Thus, make the channel somewhere between a dumb channel and a genius channel; somewhere between a channel that is just random and a channel that is all-powerful.

Lets do this by assuming that the channel is a random polynomial time adversary. Thus, Alice and Bob need only design an error-correcting code that works against a polynomial time adversary.

The key additional assumption, not too surprising, is that we will also assume that Alice and Bob can rely on the existence of a common random coin. Essentially, we are assuming that there is a pseudo-random number generator available to Alice and Bob that passes polynomial time tests.

In this model it is not hard to show the following:

Theorem 28.1. *Suppose that Alice and Bob have a code that works at a given rate for a dumb channel. Then, there is a code that works at the same rate for any reasonable channel, and further has the same encoding and decoding cost plus at most an additional polynomial time cost.*

The additional time depends on the actual pseudo-random number generator that is used; the additional cost can be near linear time.

The method is really simple. Alice operates as follows:

1. Alice creates $X = E(M)$ using the simple encoder;
2. Then she selects a random permutation π and a random pattern p. Let

$$Y = \pi(X) \oplus p.$$

3. Alice sends Y to Bob.

Than, Bob operates as follows:

1. Bob receives Y'. He computes $Z = \pi^{-1}(Y' \oplus p)$.
2. Then, he uses the decoder D to retrieve the message as $D(Z)$.

The point is that from the channel's point of view there is nothing better to do then to just pick random bits to flip. For assume the channel adds the noise N to the encoded message. Then, Bob sees $Y' = Y \oplus N$, and after he transforms it he has

$$Z = \pi^{-1}(Y' \oplus p) = \pi^{-1}(Y \oplus N \oplus p),$$

which is equal to

$$\pi^{-1}(Y \oplus p) \oplus \pi^{-1}(N) = X \oplus N'$$

where $N' = \pi^{-1}(N)$. Thus, since π is a "random" permutation, Bob can correctly decode and finally get M.

28.6 A Disaster

I made a bad mistake when I got this result, which caused me no small amount of embarrassment. I decided that the result was very cool so I sent an email to about twenty friends *with* the pdf as an attachment. Today of course I would send them a url to the paper. But in those days url's were not yet available so I sent the whole paper.

Sending this email was a bad idea. Really bad. Not everyone wants to have a large pdf file sent to them. But that was only the beginning of my disaster. The email system at the Computer Science department at Princeton went through the campus mail servers and then to the outside. The link from CS to the campus system would die while sending my email, since my message was too large for a certain buffer, being twenty times the size of a single pdf file. So it always timed-out after about 50% of the emails were sent.

You can guess what happened next. The "smart" CS code discovered the failure and so it re-sent the whole message again. And again. And again. Each time, a random subset of the emails got through.

People started to send me emails that shouted "stop already, I do not want your paper, and I certainly do not want ten copies." Some people were getting a copy every few minutes, others even more often. I quickly ran to our email experts. They said there was little they could do since the campus system was the problem. I pleaded with them to help me, and finally they got the emails to stop. All seemed well.

However, the next day I came to work, and I was getting shouting emails again. People were getting multiple copies of the paper again. I was dumbfounded. How could this be? I had not sent the message again. I felt like I was in some episode of the Twilight Zone, or in the movie "Groundhog Day."

I ran down to the email experts. Luckily, they already knew about the problem and were working to fix it. The trouble was that the CS servers had crashed that morning, and the restore was to the last "good" state, the state that was trying to send out my email. So it started the crazy problem up again. Finally, they killed off everything and no more copies of the paper went out. I never did that again.

By the way, when I presented this work at a conference, I got asked for copies of the paper from some of those who shouted back "stop." Oh well.

28.7 Open Problems

I should mention that Moti Yung and Zvi Galil worked on a version of this idea where the channel was restricted to logspace, which is still unpublished. Since the channel was only able to use logspace, they could give an unconditional proof, using Noam Nisan's famous result [112].

Suppose that we have an algorithm that needs an error correcting code. Instead of a complex code suppose we use the codes here that only work against a polynomial time adversary. What happens to the algorithm's correctness? It will not work in worst case, but will it work against a "reasonable" adversary? I do not know.

More generally can we extend the ideas here to other situations? Can we build a theory like the one developed here for channels for general algorithms?

28.8 Notes

This appeared as worst-case-complexity in the blog. Michael Mitzenmacher remarked he had used my theorem in one of his papers. It may be folklore, but he says he has never seen it written down before my paper. Adam Smith had some additional comments on the relation between the information theory community and ours. Panos Ipeirotis wondered if the ideas here could be used in game theory, which is a nice idea.

Chapter 29
Savitch's Theorem

Walter Savitch did seminal work in computational complexity theory, and has made many important contributions. For all his other wonderful work, he is undoubtedly best known for his beautiful result on the relationship between deterministic space and non-deterministic space. Probably everyone knows his famous theorem [119]:

Theorem 29.1. *For any* $s(n) \geq \log n$,

$$\mathsf{NSPACE}(s(n)) \subseteq \mathsf{DSPACE}(s(n)^2).$$

There are countless descriptions of the proof of his famous theorem: in books, on the Net, and on blogs, So if somehow you do not know the proof check them out. My goal is to talk about a missed opportunity, seeing the forest for the trees, and about the future.

29.1 Language Theory

My understanding of how Savitch came to prove his theorem is based on talking with the main players, and looking carefully at the literature. I could have some details wrong, but I think the story is too good to not be true.

Savitch was a graduate student working for Steve Cook, when he proved his theorem; actually the theorem essentially was his thesis. Where did the idea for his theorem come from? Did he create it out of thin air? Looking back from today, the argument is so simple, so elegant, so clearly from the "book" that it seems hard to believe that it was not always known. But it was not known before 1970 when he proved it.

Well he did not create it out of thin air. In the 1960's one of the main focuses of people we would call complexity theorists today, was *language theory*. This was the study of regular languages, context-free languages, and context-sensitive languages. These correspond, of course, to finite automata, non-deterministic pushdown automata, and to linear bounded automata, respectively. The rationale for this interest was driven by Noam Chomsky and Marcel-Paul Schützenberger, especially Chomsky who was interested in a theory of "natural languages". At the same time, programming languages were beginning to appear that needed a theory to support compiler construction.

Language theory was an important source of questions for computational theory to work on. One called the LBA problem, was raised in the 1960's and took thirty years to solve, as related in Chp. 18.

In 1965 at the IFIP Congress in New York City, an important paper was presented: "Memory bounds for recognition of context-free and context-sensitive languages" by Philip Lewis, Juris Hartmanis, and Richard Stearns (LHS). Another version was presented at FOCS. [68]

IFIP stands for "International Federation for Information Processing," and once was one of the top international conferences. Today there are so many other international conferences that it is probably fair to say that it is far less important. In Chp. 32, I will pointed out that I first met Dick at the 1974 IFIP Congress, so for me IFIP will always bring back good memories. When, Lewis, Hartmanis, and Stearns presented their paper, IFIP was one of the top conferences.

They sketched the proof of a number of theorems in their paper, but the one that we are concerned with was the following:

Theorem 29.2. *If L is a context-free language, then*

$$L \in \mathsf{DSPACE}(\log^2 n).$$

This is a hard theorem–I knew the proof once, and still cannot explain it without looking it up. The "miracle" is that they must show how to simulate a pushdown that can hold as many as n symbols, with only $\log^2 n$ space. This is not easy.

29.2 Cook Makes a Phone Call

What Savitch realized is that the LHS paper had a fundamental idea, an idea that was very clever, yet an idea that even the authors did not completely understand. In proving that every context-free language was accepted by only $\mathsf{DSPACE}(\log^2 n)$, they had use a clever method for keeping track of the pushdown states. Essentially, what Savitch realized is that when separated from the extra issues of context-free languages, what they were doing was exactly what he needed to prove his theorem. The paper LHS had the key idea, but by applying it to the less interesting question of space complexity for context-free languages they missed a huge opportunity. They came very close in 1965 to proving Savitch's theorem. We should not feel too bad: two of them went on to get the Turing award for other work. But they were close.

Apparently, once Savitch and his advisor Cook realized that they could use the method of the LHS paper–not the theorem itself–they called up Hartmanis. They told Juris they thought that his 1965 result was terrific, Juris told me that he was always happy to hear that they liked one of his papers. They asked some technical questions and then rang off. In a short while Hartmanis got to see their paper. He then realized what they were doing, and realized the great result that he almost got. Oh well. We cannot get them all.

29.3 A Review

In the 1972 volume 3 issue of the Journal of Symbolic Logic there is a very telling review on the LHS paper by Walter Savitch. Here is the review:

> This paper summarizes the known results on the tape complexity of context-free and context-sensitive languages. It also presents a number of new results in the area. The notions of tape complexity used are those introduced in the article reviewed above. The principal new results deal with context-free languages. A number of specific context-free languages are presented and their exact location in the tape complexity hierarchy is determined. It is shown that all context-free languages are $(\log n)^2$-tape recognizable. The proof of the $(\log n)^2$-tape bound is quite intricate, and this article gives only a sketch of the proof. The proof is, therefore, hard to read; however, the techniques are interesting and useful. The reader with perseverance will be rewarded for his effort. Walter Savitch.

I love the end of the review: *The proof is, therefore, hard to read; however, the techniques are interesting and useful. The reader with perseverance will be rewarded for his effort.*

Savitch certainly was rewarded for his effort: a thesis, and a great theorem.

29.4 Open Problems

There are two points to make. First, sometimes the proof methods of a theorem are more important than the theorem. That means that we must read the proof itself to see if the method of proof can be used elsewhere. Many times the method of proof is standard and the actual theorem is the key advance. However, as Savitch's theorem shows sometimes the method is what we must understand if we are to make further progress.

Second, one of the biggest embarrassments of all complexity theory, in my view, is the fact that Savitch's theorem has not been improved in almost 40 years. Nor has anyone proved that it is "tight." This is one of the great open questions of complexity theory.

I have thought about methods to improve his theorem. But I have no good ideas to suggest, I have no approach, I am lost.

29.5 Notes

This appeared as savitchs-theorem in the blog. Ken Regan explains the proof of Savitch's Theorem by referring to the famous saying of Confucius, "A journey of a thousand miles begins with a single step." Ken's version is, "A journey of a thousand miles must have a step when you have gone exactly 500 miles and have exactly 500 miles to go."

Chapter 30
Adaptive Sampling and Timed Adversaries

Jeff Naughton is a top expert in the area of databases. He was at Princeton, until he was stolen away to join the computer science department at Madison, Wisconsin. Jeff has a great sense of humor, is a terrific colleague, and is a great researcher.

I shall talk about two projects that Jeff and I did together. The first is a randomized algorithm that uses an *adaptive* sampling method to solve a database problem. I always liked this result, because so few randomized algorithms, that use sampling, need to be adaptive. The second is a "hole" in the worst case model of complexity theory. The hole makes certain classic theorems of complexity theory "wrong." That's right.

I enjoyed working with Jeff on both of these projects. The first, on sampling, was submitted to SIGMOD, was accepted there, and then received the best paper award. I am, therefore, 1/1/1 (submit/accept/award) at SIGMOD–I have never submitted another paper to that conference, nor will I ever. The conference was fun too, part because it was held in Atlantic City at a hotel-casino, and part because it had a magic show as the banquet entertainment.

Magic is somehow related to theory and mathematics. For example, Steven Rudich is a terrific amateur magician; Grzegorz Rozenberg, a top language theorist, is also a professional magician. Although with him please call them "illusions," never "tricks."

One thing I have learned after years of watching magic and reading about it, is the *principle of least work*. In any magic trick–excuse me, illusion–there are usually extra constraints that are placed on the magician or the assistants. A classic example, is the rug Harry Houdini used to cover the stage so you could be sure there were *no* trapdoors, in his famous illusion making an elephant disappear from a stage. The principle of least work implies that this constraint, the rug, is required to make the illusion work. It is.

I sometimes think that the key to solving a hard problem is to apply this principle. Can an obstacle be turned into the key that yields the solution to a problem? Can we use the principle of least work in our quest for results?

30.1 Transitive Closure Estimation

The first problem was estimating the size of a database query, without actually computing the query. One motivation for this, is query optimization: often a database query can be performed in different ways. If one way leads to a huge immediate answer, then it might be better to perform the query by another method.

The problem we studied is quite simple to state: Given a directed graph G, what is the size of the *transitive closure* of the graph? The transitive closure is simply the graph obtained by adding an edge from vertex x to vertex y, if there is a directed path from x to y. A graph G can have few edges, and yet its transitive closure can be much larger. A prime example, is a directed path

$$1 \to 2 \to 3 \to \cdots \to n-1 \to n.$$

This graph has $n-1$ edges. Its transitive closure has,

$$n-1+n-2+\cdots+1 \approx n^2/2$$

edges. For large n this is huge increase. What the database folks wanted to know, according to Jeff, was for a given graph, how large is the transitive closure?

Luckily, we did not need the exact answer, an approximation would be fine: we could be off by a multiplicative factor. But, the time for doing the estimate had to be as close to linear as possible. We were able to partially solve this problem [97]:

Theorem 30.1. *Suppose that G is a directed graph with n vertices. If the degree of G is bounded, then there is a randomized algorithm that in $O(n)$ time, determines the size of the transitive closure of G to within a factor of 2, with high probability.*

See our paper for the details, and for a more general trade-off between the running time and the accuracy. Later Edith Cohen proved the theorem without the restriction of bounded degree [34]. Her work is very clever, and is based on seeing the structure of the problem much more clearly than we did.

30.2 A Sampling Lemma

I will not go into detail on how we proved this theorem, but will instead explain our main lemma. Perhaps this lemma can be used in other applications.

Consider an urn that contains n balls. Each ball has a label, which range from 1 to n. The task is to estimate the sum of the labels on the balls. As you might guess–it's an urn–you may randomly remove a ball and see its label. The catch is this: the cost of sampling a ball is equal to its label.

The problem is to estimate the sum of the labels, but keep the cost low. This seems to force the sampling to be adaptive. Here is a heuristic argument that shows that using a fixed number of samples will not work.

Let Alice attempt to estimate the sum of the urn, which was created by Bob. Suppose she fixes the number of samples in advance to s. If s is large, then Bob will make all the balls have size n, and her cost will be sn, which is a non-linear cost. On the other hand, if s is small, then Bob can do two things. On the one hand, he could make the urn all 1's. Or he could make the urn contain order $\frac{n}{st}$ balls with label n and the rest with label 1. Then, with probability,

$$(1 - \frac{1}{st})^s \approx 1 - 1/t$$

Alice will see only balls with the label 1, but, the sum of the labels is $\Omega(n^2/st)$. Alice seems to be stuck: if she says the urn sum is about n, then she can be wrong; if she guesses it is larger, then also she can be wrong.

There is a simple adaptive sampling algorithm that does work. Randomly sample, with replacement, balls with labels $b_1, \ldots, b_k$ until

$$b = b_1 + b_2 + \cdots + b_k \geq cn$$

where $c > 0$ is a constant. Then, stop and estimate the sum of the urn as

$$v = \frac{bn}{k}.$$

This is clearly an unbiased estimator for the sum of the balls in the urn. The key issue is whether the estimation v is likely to be good.

Lemma 30.1. *The value v is likely to be within a factor of 2 of the urn sum, for c large enough.*

The exact relationship between c and the probability is in our paper. Note, the method always has cost at most linear.

30.3 Timed Adversaries

One day, when Jeff was still at Princeton, he pointed out to me a curious fact: in the last few years, he said, no assistant professor had ever been hired to the department, *when I attended their job talk.* For example, I missed Jeff's. I do like to ask questions, and sometimes can get into trouble, but his comment bothered me. I did once ask a question at a system job talk that got the answer: "that is a dangerous question." But, that's another story. Jeff's comment bothered me, so my solution was to attend all the job talks the following year, every single one. Since we did hire that year, the string was broken.

The second question arose from a job talk I attended with Jeff at Princeton–okay I did sometimes cause problems. The speaker was talking about a network routing problem, and he stated a theorem. The theorem was of the form:

Theorem 30.2. *Against any adversary, with high probability, all routing steps will take at most $O(\log n)$ time.*

This is the outline of the theorem, but as he stated it, I realized that it was "false." There were adversaries that could defeat the algorithm. Yet his theorem was technically correct.

Confused? The issue is a fundamental assumption built into the adversary worst case model, that is usually true. But, in his application, I thought that it could easily be false, since he allowed the routing requests to be created by potentially adversarial programs. After his talk, Jeff and I started to discuss the issue and we soon realized that there was a problem in his worst case model: it was not worst case enough.

30.4 The Attack

The issue is clocks. Algorithms can have clocks, and therefore algorithms can tell time. This ability is missing in the usual model, but is always there in practice. Instead of explaining the problem with the job candidate's result, I will explain it in the simpler context of hashing with linear chaining.

This is a basic and popular data structure. Recall that there is a hash function that selects where to place an object, and if that location is taken the object is put into a linear list. To lookup an object, the hash function is used to find the linear list, and then that list is searched.

This is a popular and useful data structure. Both the insertion and lookup operation take expected time $O(1)$ time provided the number of objects is not too large compared with the range of the hash function. However, if we assume that an adversary supplies the values to be stored, then it is possible to make the behavior of the data structure dramatically worse. The adversary need only arrange that all the values hash to the same location.

The standard answer to this issue, in both theory and practice, is to select the hash function randomly from a family of functions. Then, it can be proved that with high probability the hash table operates as expected: operations take constant time.

This is essentially what was claimed in the routing theorem, by our job candidate. What Jeff and I showed is that this is wrong [97]. Provided the hash function family was known there was an attack that could make the data structure perform badly. The attack was based on the ability to time operations.

The attack works in two phases. In the first phase the exact hash function is discovered by using timing. In the second phase this is used to flood the data structure with values that are stored at the same location.

The first phase is simple: start with an empty data structure and begin adding randomly selected x's to the data structure. Suppose that we had a way to discover x and y that collide, i.e. so that

$$h_j(x) = h_j(y)$$

where $h_j(x)$ is the hash function. Then, we show that we can unravel the value of j. Note, this works for a large class of popular hash functions. For others, several values may be needed; see the paper for details.

The only issue is, how do we find collisions? Since the values are random, collisions will occur by the "birthday paradox" in about $\sqrt{m}$ operations, where m is the range of the hash functions. Thus, the only issue is how do we detect that a collision has occurred? The attack uses the ability to clock operations to discover which are collisions and which are not. Simply, collisions yield longer lists, which take more time. If you are worried about how accurate the timing must be, then we can repeat operations to get a better statistical sample. But the idea is sound, and real implementations of this attack do occur.

30.5 Time in Cryptography

Timing attacks have been used in cryptography to break systems, that would otherwise be unbreakable. The extra information available via timing often suffices to break very powerful systems.

Paul Kocher first showed how to use timing to break certain implementations of RSA, while Dan Boneh and David Brumley later showed how to attack OpenSSL– an important practical security protocol [82, 24]. This latter work is especially interesting, since Dan and David can do their attack on servers running remotely.

It is interesting to point out that time can be used in real system attacks. Jeff and I were the first to publish the potential of using time for breaking systems, but it seems clear that Paul's work was completely independent of ours.

30.6 Open Problems

One open problem is to determine the exact size of the transitive closure of a graph in linear time. While the random approximation algorithms are quite useful, there is no reason I know that precludes the possibility that there is a linear time *exact* algorithm. Note, for undirected graphs the problem is easy to do in linear time: find the connected components in linear time, and then compute the size of the transitive closure.

Can the techniques of property testing be applied to estimate the size of the transitive closure? Or are there other ways to estimate the size of the transitive closure? Note, the estimation of the transitive closure of a directed graph is closely related to a matrix problem. Suppose that A is an n by n sparse invertible matrix. How many non-zero entries will A^{-1} have? Perhaps this connection can be used with known low rank approximation methods to beat our bounds.

Finally, the "lower bound" I sketched to motivate adaptive sampling is not a rigorous theorem. Make it one. What are the best bounds possible for this urn problem? Or even for generalizations where balls can have general labels?

30.7 Notes

This appeared as adaptive-sampling-and-timed-adversaries in the blog. Pascal Koiran asked a nice question: is a directed acyclic graph easy for transitive closure? The answer is no—it can be proved that this is actually the hardest case.

Chapter 31
On The Intersection of Finite Automata

George Karakostas and Anastasios Viglas are two excellent theorists, who both have their Ph.D.'s from the Princeton computer science department. I was the Ph.D. advisor of Viglas, and later had the pleasure of working with both of them on several projects.

I want to talk about finite state automata, and an open question concerning them. The question is very annoying, since we have no idea how to improve a classic algorithm. This algorithm is decades old, but it is still the best known. There seems to be no hope to show the algorithm is optimal, yet there also seems to be no idea how to beat it. Oh well, we have seen this situation before.

It was a great pleasure to work with George and Anastasios. I enjoyed our work together very much, and they have continued working together over the years—on a wide variety of topics.

I do not have any really great stories about them, or even one of them. One story that does comes to mind is about steaks. George likes his steak cooked well done, as I do too. I know that is probably not the proper way to order a steak, but that is the way I like them. But, when George says "well-done" he means "well-done." I always loved the way George would explain this to our server: They would ask George, "how would you like your steak cooked?" And George would say, "well done—actually burn it." This usually caused some discussion with the server, and George had to be insistent that he really wanted it "burnt." I like someone who is definite.

Let's turn from steaks to finite automata, and to a question about them that is wide open.

31.1 Intersection of Finite State Automata

The power of deterministic finite state automata (FSA) is one of the great paradoxes, in my opinion, of computer science. FSA are just about the simplest possible computational device that you can imagine. They have finite storage, they read their input once, and they are deterministic—what could be simpler? FSA are closed under just about any operation; almost any questions concerning their behavior is easy to decide. They seem "trivial."

On the other hand, FSA are more powerful than you might guess—this is the paradox. While FSA cannot even count, they can do magic. FSA play a major role in BDD's, they can be used to prove the decidability of complex logical theories such as *Presburger's Arithmetic*, and they are used everyday by millions. Every time you search for a pattern in a text string, especially when you use "don't cares", the search is done via FSA technology. Simple, yet powerful.

I want to raise a question about FSA that seems to be interesting, yet is unyielding to our attempts to understand it. In order to understand the question we first need some simple definitions. If A is a FSA, let $L(A)$ be the language that it accepts, and let $|A|$ be the number of states of A.

One of the classic results in the theory of FSA is this:

Theorem 31.1. *If A and B are FSA's, then there is an FSA with at most $|A| \times |B|$ states that accepts $L(A) \cap L(B)$.*

The proof of this uses the classic Cartesian product method: one creates a machine that has pairs of states, and simulates both A and B at the same time. This is why the simulation uses $|A| \times |B|$ states. A corollary of this result is the following:

Corollary 31.1. *If A and B are FSA's, then there is an algorithm that runs in time $O(|A| \times |B|)$ and decides whether or not $L(A) \cap L(B)$ is empty.*

George, Anastasios, and I noticed that there is no reason to believe that this cannot be improved. I guess that is a double negative, what I mean is: let's find an algorithm that solves the emptiness problem for $L(A) \cap L(B)$ in time linear in the description the two FSA's. Note, there are FSA's A and B so that the smallest number of states of a machine C so that

$$L(C) = L(A) \cap L(B)$$

is approximately the product of the number of states of A and B. Thus, a faster algorithm must avoid the step of building the product automata. But, this is precisely the issue I have raised before: the algorithm for determining emptiness is free to do anything it wants.

31.2 Our Results

We worked hard on trying to beat the standard approach, yet could not improve on it. So eventually we wrote a paper that in the true spirit of theory, assumed that there was a faster than product algorithm for the intersection of FSA [75]. Then, we looked at what the consequences would be of such an algorithm. Our main discovery was that if there existed faster methods for the intersection of automata, then cool things would happen to many long standing open problems.

You could of course take this two ways: you could use these results to say it is unlikely that any algorithm can beat the product algorithm, or you could use these results to say how wonderful a better algorithm would be. It's your choice. I like to be positive, so I tend to hope that there is a faster algorithm.

Suppose that $A_1, \ldots, A_k$ are FSA each with at most n states. We asked, what would the consequences be were there an algorithm that ran in time $n^{o(k)}$ and determines whether or not there was an input they all accepted. That is, the algorithm determines whether or not the following language is empty

$$L(A_1) \cap \cdots \cap L(A_k).$$

Call this the *product emptiness problem* (PEP). Note, if k is allowed to be arbitrary, then the answer is known. Dexter Kozen proved long ago that in this case the problem is PSPACE complete [84]. Thus, we have to restrict ourselves to fixed k, or allow algorithms that run in time

$$2^k n^{\sqrt{k}},$$

for example.

Here are a couple of results from our paper that should help give you a flavor of the power of assuming that PEP could be solved more efficiently than the product algorithm. As usual please check out the paper for the details.

Let F be the assumption that for any fixed k there is an algorithm for solving PEP with k FSA'a where each has at most n states in time $O(n^{\varepsilon k})$, for any $\varepsilon > 0$.

Theorem 31.2. *If $\mathscr{F}$, then for any $\varepsilon > 0$ there is an algorithm for the knapsack problem that runs in time $2^{\varepsilon n + O(1)}$.*

Theorem 31.3. *If $\mathscr{F}$, then for any $\varepsilon > 0$ there is an algorithm for factoring that runs in time $2^{\varepsilon n + O(1)}$.*

Theorem 31.4. *If $\mathscr{F}$, then for any $\varepsilon > 0$,*

$$\mathrm{NDTIME}(n) \subseteq \mathrm{DTIME}(2^{\varepsilon n}).$$

31.3 A Sample Proof

None of the proofs of these theorems is too difficult, but the most complex is the last result. This result uses the block-respecting method [66]. Here is a sample proof that should demonstrate how $\mathscr{F}$ is used. I will outline the proof of the factoring theorem, proving a $2^{2n/3}$ type bound.

Let M_p be the FSA that given the input x, y, z checks whether or not

$$x \times y \equiv z \bmod p$$

and neither x nor y is 1. We assume that x and y are n bit numbers and z is a fixed $2n$ bit number that we wish to factor. Then select three distinct primes p, q, r each about size $2^{2n/3}$. Look at the PEP,

$$L(M_p) \cap L(M_q) \cap L(M_r).$$

There are two cases. If there are no solutions, then z must be a prime. If there are solutions, then suppose that x, y, z is one. Then,

$$x \times y \equiv z \bmod pqr.$$

But, $pqr > 2^{2n}$ so that this is an equality, and so $x \times y = z$.

The above is a primality test. We make it into a factoring algorithm by adding a simple test: we insist that the first automaton also test that x lie in some range. This requires only a polynomial factor increase in the number of states of M_p. But, we can then do a binary search to find a factor of z.

31.4 Open Problems

Can one solve the PEP in time faster than the product algorithm? Even for two automata this seems like a neat problem. I find it hard to believe that the best one can do is to build the product machine, and then solve the emptiness problem.

I do have one idea on a possible method to improve the PEP in the case of two automata. Suppose that A and B are two FSA. Then, construct the non-deterministic pushdown automaton that operates as follows on the input x: The machine reads and pushes the input onto the pushdown store; at the same time checking whether or not the input is accepted by A. Then the machine pops off the input from the pushdown, and checks whether or not the value coming off the pushdown, x^R (the reversal of x), is accepted by the machine B.

This can be done by a non-deterministic pushdown automaton that uses at most $|A| + |B| + O(1)$ states. The emptiness problem for pushdown machines is in polynomial time. If it could be done is less than quadratic time, then at least for two automata we would be able to beat the product algorithm. An open problem is, can this be made to work?

31.5 Notes

This appeared as on-the-intersection-of-finite-automata in the blog. There were a number of comments. Moshe Vardi pointed out that automata theory was once the main area of theory, and it still contains many open problems.Daniel Marx connected the ideas here to parameterized complexity.

Chapter 32
Where are the Movies?

Dick Karp and I first met at the IFIP Congress in Stockholm in August of 1974. I recall the meeting vividly. I was, then, a freshly minted assistant professor at Yale Computer Science Department, when I met him. Dick was quite friendly and gracious towards me and I have ever since considered him a close friend. Actually the truth is I was a "Gibbs Instructor." I had a reduced teaching load, only a two year contract, and a small travel budget. I also was paid 10 percent less than my colleagues—David Dobkin and Larry Snyder—who were real assistant professors.

Another reason I recall the time so clearly was that Richard Nixon resigned as the 37th president of the United States on August 8th. I watched, from my hotel room, a Swedish TV announcer say that Nixon had "avången." I know no Swedish, but I could see that this looked like "Ausgang" which is exit in German. Great, Nixon was gone.

Back to the present, well at least this century. A few years ago I gave a talk at the opening of a new research center at Georgia Tech. The center is called ARC and stands for "Algorithms and Randomness Center." It was created by Santosh Vempala and has been a terrific success. We often call it the ARC center which is of course a bit redundant since that translates into "Algorithms and Randomness Center Center," but that's the way it goes.

My talk was on "Algorithms are Tiny" which I discussed in Chp.2. The point of the talk was that algorithms are like equations. They are small, yet can have huge impact. They are small, yet can create entire new industries. They are small, yet can change the world. To paraphrase Einstein: "Algorithms (he said equations) are more important to me, because politics is for the present, but an algorithm is something for eternity." Just as $E = mc^2$ changed the world, I argued that there could be a one page long algorithm for SAT or for factoring. Such a single page could change the world.

At the end of the talk I got the usual technical questions. Finally, Dick raised his hand and I called on him. He said if I was right about algorithms being "tiny," then where were the movies? Where were the movies about the discovery of a new algorithm that changes the world. Indeed where are they?

The closest there has been is probably the 1992 film "Sneakers" which starred Robert Redford. Len Adleman was the technical advisor and worked not for money, but provided he and his wife could meet Redford.

32.1 Open Problems

What a great question. I really had no good answer. Shouldn't there be some cool sci-fi type movie on a story like this:

A computer scientist discovers how to factor. She gets into trouble with various people who either want to suppress the discovery, or to get it to break commercial systems to make money, or others who want to break military systems to attack the US. And there would also be a love story between her and some agent...

Perhaps I should stop working in theory for a while and work on a screenplay instead. My daughter Andrea lives in LA and she knows people who know people; so I could get my story to the right people.

32.2 Notes

This appeared as where-are-the-movies-on-pnp in the blog. Martin Schwarz pointed to the movie **Pi** as an example. Several people liked the TV show **Numb3rs**, and said it often had deep mathematical themes. John Sidles pointed to Norbert Weiner's novel "The Tempter," and said he did not think it ever became a movie.

Part III
On Integer Factoring

Chapter 33
Factoring and Factorials

Adrien-Marie Legendre was one of the great number theorists of the 19th century. He proved a whole host of amazing results. One wonders what he would think of the applications and consequences of his work over a century later. One of his great discoveries is the *Duplication Formula* for the *Gamma Function* [10]. The Gamma function, $\Gamma(z)$, is a function defined for all complex numbers that agrees with the factorial function at the natural numbers. In a sense you should think of the Gamma function as a generalization of the factorial function. The factorial function is, of course, defined for natural numbers n as:

$$n! = 1 \times 2 \times 3 \times \cdots \times n.$$

The exact relationship between them is that $\Gamma(n+1) = n!$. You might ask, why the $n+1$ and not n? There is an interesting story here, but I am not the right one to tell it, so for now we will just have to accept it.

So what is the duplication formula, and why is it connected to factoring? I will explain. The formula is this: for all complex z,

$$\Gamma(z)\Gamma(z+\frac{1}{2}) = 2^{1-2z}\sqrt{\pi}\Gamma(2z).$$

The rationale for calling it the "Duplication Formula" is that the left hand side $\Gamma(z)\Gamma(z+1/2)$ is almost a "duplication". Except for the $1/2$ the formula would have exactly two occurrences of the Gamma function at the same value. Hence, the name the "Duplication Formula." However, the two Gamma functions are evaluated at slightly different values: one at z and one at $z+1/2$. This small difference is not, in my opinion, exactly a duplication: perhaps it should be called the "Almost Duplication Formula." My guess is that this name would not be nearly as cool and exciting.

So what is the connection between the Duplication Formula and factoring? We will see shortly that if the formula were a *real* duplication formula, then we could factor in polynomial time. Thus, Legendre came within a $1/2$ of having discovered a formula that would be able to factor fast, would break modern factoring based crypto-systems like RSA, and generally would be a huge result. Alas, there is that small but critical $1/2$. It barely misses. As agent Maxwell Smart on the old TV show "Get Smart" used to say, "missed by that much." More precisely, if the formula was instead

$$\Gamma(z)\Gamma(z) = c2^w \Gamma(2z)$$

for some constant c and some w a polynomial in z, then factoring would be easy. The rest of this post is an explanation of why this is true.

33.1 Factoring with Factorials

One of the curious properties of $n!$ is a folk theorem about its arithmetic complexity. Say that a function $f(n)$ has arithmetic complexity $t(n)$ provided there is an arithmetic circuit that computes $f(n)$ in $t(n)$ steps. As usual an arithmetic circuit starts with the variables and the value 1 , then at each step it performs one of the basic arithmetic operations from $\{+, \times, \div\}$ on a pair of previous values, including the constant -1. The answer is, of course, the last value computed.

One of the major assumptions of modern cryptography is that factoring of integers is hard. For cryptography one usually needs a less general assumption: Suppose that $X = pq$ for p and q random primes in a range $[2^m, 2^{m+1}]$. Then it is presumed hard to find p and q from just X provided m is large enough. Note, often there are additional constraints placed on the primes, but these will not affect our discussion.

By "hard" we can mean that there is no polynomial time algorithm, or even that there is no circuit of polynomial size. The problem with this assumption is many-fold. First, the problem of factoring is in $\mathsf{NP} \cap \mathsf{co\text{-}NP}$ so it is unlikely to be NP-complete. Second, the best algorithms for factoring are "sub-exponential." For example, the approximate running time of the Number Field Sieve [88] is:

$$\exp(c \log^{1/3} X)$$

where c is about 2. (There are some much smaller terms in the exponent, but $\log^{1/3} X$ is the main term.) Thus, if factoring were NP-hard, there would be dire consequences for complexity theory. I will just remark that at a conference a few years ago Avi Wigderson spoke on the uses of the factoring assumption, and Michael Rabin spoke on the possibilities of how factoring could be easy.

I find, like many others, this problem to be extremely interesting. I want to point out an old but interesting connection between factoring and factorials. . Here is the folk theorem:

Theorem 33.1. *If $n!$ can be computed by a straight-line arithmetic computation in $O(\log^c n)$ steps, then factoring has polynomial size circuits.*

We have stated the theorem as a non-uniform result. If you want a uniform bound then you need to make the stronger assumption that the arithmetic circuit for $n!$ is uniform. That is, for each n there not only is a straight-line computation, but what operations to do at each step is computable by a polynomial time algorithm.

The proof of this folk theorem is simple. Suppose that you wish to factor some non-zero number X that is composite. For any number $1 \leq y \leq X$ define the number $t(y)$ as $\gcd(y!, X)$. (As usual $\gcd(a, b)$ is the greatest common divisor of a and b.) We claim that $t(y)$ has the following key property: either $t(y) = 1$ or $t(y) = X$ or $t(y)$ is a proper factor of X. This follows directly from the definition of the gcd.

Now we will use this function $t(y)$ and binary search to get a proper factor of X. Once we have such a factor we can get all the factors by applying the method again. The binary search works as follows: set $a = 1$ and $b = X - 1$. Note, that the interval $[a, b]$ has the following property: $\gcd(a!, X) = 1$ and that $\gcd(b!, X) = X$. The latter follows since $\gcd(b!, X) = \gcd((X - 1)!, X)$ cannot be 1; thus, if it is less than X we have a proper factor. In general we will always have an interval $[a, b]$ so that $\gcd(a!, X) = 1$ and $\gcd(b!, X) = X$.

Suppose we have such an interval. We claim that $a + 1$ cannot equal b. Assume by way of contradiction that $a + 1 = b$. Since $a = b - 1$ it follows that $\gcd((b - 1)!, X) = 1$ and $\gcd(b!, X) = X$. Thus, b must equal X which is impossible. So $a + 1 < b$. Let c be the number that is closest to the mid-point between a and b. Check whether or not $\gcd(c!, X)$ is a proper factor of X. If it is not then there are only two cases. In the first the gcd is 1. In this case go to the interval $[c, b]$. In the second case the gcd is X, in this case go to the interval $[a, c]$. Since this process stops eventually it must find a proper factor of X.

There is one missing, but critical point. We cannot afford to compute the factorials used in the above procedure. They will be too big. So we use the arithmetic circuit to do all the above computations, but do all of them modulo X. The key insight is that this does not change any of the gcd calculations. This depends on the simple fact that $\gcd(z, X)$ is the same as $\gcd(z \bmod X, X)$.

33.2 What is The Complexity of Factorials?

So what is the complexity of computing the factorial function?. Since currently factoring is believed to be hard, clearly $n!$ must be hard to compute. But is it really hard?

Let's recall the Duplication Formula:

$$\Gamma(z)\Gamma(z + 1/2) = 2^{1-2z}\sqrt{\pi}\Gamma(2z).$$

If the formula were instead

$$\Gamma(z)\Gamma(z) = c2^w\Gamma(2z)$$

then there would be a fast algorithm for factorial. Then we would be able to factor. Let $G(n)$ be the cost of computing $\Gamma(n)$. Then, such a equation would yield that

$$G(n) \leq G(n/2) + O(\log^{O(1)} n).$$

This implies a fast method of computing $n!$.

Imagine that

$$\Gamma(2z) = R(z, \Gamma(z), a_1, \ldots, a_l) \tag{33.1}$$

where $R()$ is a rational function and each a_i is an exponential term. That is a term of form $b(z)^{c(z)}$ where b and c are polynomials. If (33.1) was true, then $n!$ would be easy to compute. Note, this remains true even if the identity only holds true for a dense subset of the natural numbers S. That is S must satisfy the property that for each x there is a $y \in S$ so that $|x - y|$ is at most $O(\log^{O(1)} x)$.

Furthermore, suppose that $\Gamma^*(n)$ is a function so that

$$\gcd(\Gamma^*(n), n!) = n!$$

for all natural numbers n. Then computing this function fast would be enough. Note, the point of this generalization is that you cannot as easily show that such a formula does not hold because of some growth argument.

33.3 Open Problems

Can we prove that $n!$ cannot satisfy any such equation? I think that this might be hard, but it would certainly be interesting to have such a result. While it would not be a general lower bound, having a result that there is no real "duplication formula" would at least one approach to factoring must fail. On the other hand having no such lower bound shows how little we know about factoring. Those who believe strongly that factoring is hard should be worried that we cannot even dismiss the possibility that such a factoring method may exist.

Another direction is discussed in [95]. This approach is to look not at the factorial function directly but indirectly via a polynomial. Consider the polynomial $f_n(x) = (x-1)(x-2)(x-3)\ldots(x-n)$ defined for each n. If we can compute this polynomial in s steps, then clearly we can compute $n!$ in s steps: just set $x = 0$. However, suppose more generally that $g_n(x)$ is a polynomial with n distinct integer roots. If we can compute $g_n(x)$ fast then can we still factor fast? The answer is yes under mild conditions on the distinct roots.

33.4 Notes

This appeared as factoring-and-factorials in the blog. Neil Dickson had an interesting comment on new equations for the Gamma function. Aram Harrow asked a very cool question, "would a factoring algorithm imply a fast algorithm for $n! \bmod p$?"

Chapter 34
BDD's

Randy Bryant is famous for his work on BDD's, which are one of the most powerful tools used today in the design and analysis of digital logic. While Randy has done other great work, his paper [25] on BDD's created an avalanche of research, papers, systems, and applications. This famous paper is one of the most cited in all of computer science. There are many additional papers that should probably cite him–after a while you stop getting citations, since the work is so well known. This phenomena is related to *Ron Graham's Award Lemma*:

Lemma 34.1. *For every field of science with n awards, there is a $k = k(n)$, so that if you win at least k awards, then you will win at least $n - k$.*

Let's talk about BDD's from the view of complexity theory, not from the view of a logic designer. I will explain what they are, why they are important, and the relation to problems near and dear to me–like factoring. Always factoring.

At the 50th anniversary celebration of the creation of the School of Computer Science at Carnegie-Mellon (CMU) Randy, as Dean, made a comment that I think sums up his approach to many things. The 50th, in 2006, was a big event, and one of the greats of AI was speaking to a packed hall. Ed Feigenbaum was talking about the seminal work that was done at CMU, in the early years, especially highlighting the work of Allen Newell and Herbert Simon on list processing. Now there is no doubt that they were the first to create list processing languages, but it is unclear whether they actually invented the concept of a list. In any event, they shared a well deserved, Turing award in 1975 for this and related work.

When Ed finished his talk to huge applause, Randy, who was in the audience, stood up and started to question whether Newell and Simon actually invented the notion of a list. John was definitely taken aback: why would the Dean ask these questions about two of CMU's greatest scientists? But, Randy is a true independent thinker, and he wanted to know–in front of hundreds of people–did Newell and Simon actually invent the list concept or not? The session chair quickly stepped in and the celebration moved on to another talk.

This was an especially interesting debate, since Newell was one of the wisest and kindest faculty that CMU probably has ever had; truly he is beloved by the entire

CMU community. He taught me what little AI I know–in those days long ago AI was called Complex Information Processing (CIP).

I still remember his great sense of humor, which helped make him such a wonderful teacher. The first day of his CIP class, our classroom was completely full so that at least twenty students had to stand in the back of room. They had no seats, no desks. One of them raised a hand and asked Newell if we could possibly move to a larger room. He simply said, "in my experience, the room will grow in size as the class goes on." Of course he was correct: after a few more class meetings, there were plenty of seats.

Randy never did get an answer to his question: who invented lists? I think raising the question shows his quest for knowledge, his ability to think out of the box, and his ability to not care what the answer is–as long as it is correct. I respect him very much for these traits.

34.1 BDD's

So what is a BDD? It is a *binary decision diagram*. Simply put, it is a way to represent a boolean function that allows the discovery of properties of boolean functions that would otherwise be hard to determine.

Actually, Randy studied what are called *Ordered Binary Decision Diagrams* (OBDD), and today what I call BDD's are a special case of OBDD's–confused? I am doing this for two reasons. The special case that we will study has, up to a small factor, the same size as the more general OBDD's. So if you are an expert, please do not get too upset at me; and if you are not, just be aware that I have simplified things to avoid complex notation. The essential ideas, and the power of the method, are still the same.

Also by restricting ourselves to this special, but powerful class of BDD, we will be able to extend the notion of BDD's in an interesting way. Since this extension is more interesting to theorists than to logic designers, I believe that it is not yet well explored. But, I think it has great potential.

One of the great mysteries to me about BDD's is that they are not–I believe– used much in the complexity theory arena, yet they are a simple generalization of something that we know well: finite state automata (FSA). One way to view a BDD is as a kind of finite state automaton on "steroids."

Suppose that $f : \mathscr{B}^n \to \mathscr{B}$ where $\mathscr{B} = \{0, 1\}$. In complexity theory we have many ways to define the complexity of f. What BDD's do is define the complexity of f as this: Let $\lambda(f, \pi)$ be the size, in states, of the smallest deterministic finite state automata that accepts the language:

$$L = \{x_{\pi_1} x_{\pi_2} \ldots x_{\pi_n} \mid f(x_1 x_2 \ldots x_n) = 1\}$$

where π is a permutation. Then, define $\mathrm{BDD}(f)$ as the minimum of $\lambda(f, \pi)$ over all permutations π.

A classic example, demonstrating the power of BDD's, is the language of all strings of the form

$$x_1, \ldots x_m y_1 \ldots y_m$$

where $x_i = y_i$, for all indices i. It is easy to see that an FSA for this language requires 2^m states: after scanning the first half of the string, the automata must "remember" 2^m different values. Any fewer number of states, forces the automata to confuse two strings, and therefore make a mistake.

Yet, there is something unsettling about this example. It only works, because the string is represented in a most inconvenient manner. Suppose instead the string is represented as:

$$x_1 y_1 x_2 y_2 \ldots x_m y_m.$$

Then, the FSA needs only $O(1)$ states to accept the new language. This huge improvement is the fundamental idea of BDD's:

Allow the inputs to be re-ordered in the most convenient way.

34.2 A Basic Property of BDD's

One of the most important properties of a BDD is:

Theorem 34.1. *Suppose that* BDD(f) *is at most S and we know the permutation* π. *Then, in polynomial time in S, we can determine whether or not there is an x so that* $f(x) = 1$. *Even more, we can count the number of x such that* $f(x) = 1$ *in the same time bound.*

In many representations deciding if there is an x such that $f(x) = 1$ is NP-hard. Here we can even count the exact number of such inputs.

The proof of this theorem depends on a simple property of deterministic finite state automata. Consider an FSA M with S states, for which we wish to count the number of inputs of length n that it accepts. We create a new FSA M' with at most $n \times S$ states such that, (i) M' simulates M on length n inputs, and (ii) there are no cycles in M''s finite state diagram. Then, we can label, inductively, the states of M' with the number of inputs that can reach that state. In this manner we can count the number of inputs that M accepts.

The above argument is quite useful, and is only a slight extension of the well known fact that deciding if an FSA accepts some input is in polynomial time. Yet, this extension does not seem to be as well known as it should be.

34.3 Bounded Width Computations

There is a relation between read-once bounded width programs and BDD's. Are they not the same? The answer is both yes and no.

A BDD can be viewed as a read-once bounded width computation, where the width is bounded by the number of states. So the answer is yes.

This is a bit misleading, however. The power of BDD's is the ability to select the order of the inputs to the computation. The ability to vary the order of inputs to make the width of the computation smaller is what the theory of BDD representation is all about. So the answer is no.

Also in the next two sections, I will present two extensions to the basic BDD concept that will diverge greatly from bounded width computations.

34.4 Pushdown BDD's

There is a further reason that BDD are different from read-once computations. It is possible to generalize the BDD notion and replace finite state automata by non-deterministic push down automata. Now the complexity of a boolean function f is determined by the size, in states, of the smallest pushdown automaton that accepts the language:

$$L = \{x_{\pi_1} x_{\pi_2} \ldots x_{\pi_n} \mid f(x_1 x_2 \ldots x_n) = 1\}$$

where π is a permutation. Again we are free to chose the "best" permutation. Let us denote this minimum by BDDPUSH(f).

The reason this is interesting is that the power of a pushdown trivially allows a kind of read-twice. Suppose the input is $x = x_1 \ldots x_n$. Then by placing it on the pushdown and popping it off at the end operation, such a machine can simulate the ability to see the string:

$$x_1 \ldots x_n x_n \ldots x_1.$$

This is a special kind of read twice ability.

The counterpart of the basic theorem now is weaker:

Theorem 34.2. *Suppose that* BDDPUSH(f) *is at most S and we know the permutation* π. *Then, in polynomial time in S, we can determine whether or not there is an* x *so that* $f(x) = 1$.

Note, we do not preserve the ability, to my knowledge, to be able to count the number of accepting paths.

The proof of this theorem is simple, but a bit curious. The proof is simply to note that to determine whether $f(x) = 1$ for some x, reduces to the emptiness problem for a pushdown machine. This can be done in polynomial time in the number of states of the machine.

However, the classic literature on language theory does not seem to prove this. The literature does prove that any such language is in polynomial time, but that is not exactly what we need. It is an issue of uniformity: we need given a pushdown machine M, in time polynomial in the size of M, we can tell if $L(M)$ is empty. That is different from the well known fact that $L(M)$ is in polynomial time. Do you see the difference?

My guess is that most of the theorems about pushdown machines were proved long before the great interest in polynomial time. This is an aside, but clearly as our field advances we must re-examine old theorems and results–many may need to be re-proved in light of new concerns.

34.5 Another Extension of BDD's

Define an *encoding* as a mapping $e : B^n \to B^n$ so that (i) e is computable in polynomial time and (ii) e^{-1} exists and is also computable in polynomial time. Obviously, this means that the mapping e is an efficiently computable bijection on the set of length n bit strings.

For many applications of the BDD concept to theory, we can allow much more powerful encodings than just permutations. Clearly, a permutation is an encoding in the sense that we mean here. But, many other natural mappings could be used:

- The encoding could be an arbitrary linear invertible map over $\mathbb{Z}/2\mathbb{Z}$.
- The encoding could be a permutation polynomial over the input;
- The encoding could be a piecewise linear map.

The latter could be a map that is defined, for example, as one bijection on the interval $[0, m]$ to itself, and another on $[m + 1, 2^n - 1]$ to itself.

Of course, the more powerful the encoding allowed the more difficult it will be to prove lower bounds, or lower bounds may not even exist. Thus, it may be possible to change the complexity of a function dramatically.

34.6 Lower Bounds on BDD's

One of the first theorems that Randy proved about BDD's was a lower bound on $\mathrm{BDD}(\mathrm{mult}_n)$. Let mult_n be the boolean function that is 1 on strings of the form:

$$x_1 \ldots x_n y_1 \ldots y_n z_1 \ldots z_{2n}$$

where $x \times y = z$ as integers.

The importance of this function to Randy was that "real" digital logic had multipliers. Thus, he was quite interested in how large is the BDD representation. He proved the following pretty theorem:

Theorem 34.3. *For all n,* $\mathrm{BDD}(\mathrm{mult}_n) \geq \Omega(2^{n/8})$.

Randy used a lemma from an earlier paper [99] of Bob Sedgewick and myself on related lower bounds via communication complexity—see Chp. 27. For the details of how Randy proved this theorem see his paper [25].

Philipp Woelfel, in 2005 [136], improved this greatly to,

Theorem 34.4. *For all n,* $\mathrm{BDD}(\mathrm{mult}_n) \geq \Omega(2^{n/2})$.

These lower bounds do not apply to the extensions that we have raised, namely BDDPUSH and other types of encodings.

34.7 Factoring

The interest that I have in the previous lower bounds is their relationship to factoring. If mult_n has BDD complexity S, then we can factor numbers of length n in time polynomial in S. The same is true, if we replace the BDD complexity by BDDPUSH complexity.

Suppose that for some ordering of x, y, z there is an FSA M that accepts if and only if $x \times y = z$. Then there is an oracle $\mathcal{O}$ that can answer any query of the form: Given a, b, c, are there r, s so that $ra \times sb = c$? Here uv is the concatenation of u and v, not their product. The existence of this oracle follows directly from the main theorem on BDD's.

Then, by a standard self-reduction argument, we can factor. Suppose that we wish to factor N. We proceed to try and find the value of x so that $f(x, y, N) = 1$ for some $1 < y < N$. We do this by guessing one bit of x at a time, while restricting y.

34.8 Open Problems

There are a number of interesting open problems. What is the value of

$$\mathrm{BDDPUSH}(\mathrm{mult}_n)?$$

If it is of the form 2^{cn}, then what is the value of c? If c is small, then this could be a threat to factoring-based cryptography.

Suppose we have a boolean function f, how well can we approximate it with another boolean function g so that $\mathrm{BDD}(g) \ll \mathrm{BDD}(f)$? For example, is the multiplication function well approximated by such a boolean function?

Also what happens for non-standard encodings? Can we still prove exponential lower bounds on mult_n, or do these become too hard to prove? Of course, a better upper bound would be even more exciting.

The BDD measure of complexity is markedly different from some of our standard measures. For example, some of our favorite boolean functions, old friends, that have high complexity in other measures are trivial in this model. For example, computing parity, or more generally any threshold function is easy in this model.

34.9 Notes

This appeared as bdds-and-factoring in the blog. Moshe Vardi had a number of interesting comments. He agreed that BDD's really are essentially FSA's; he said he teaches this in his logic class at Rice. He also added the key point that finding a "good variable order is one of the most difficult challenges for BDD users."

Chapter 35
Factoring and Fermat

Pierre de Fermat is perhaps best known for the proof that he never wrote down, and perhaps never had.

> I have discovered a truly remarkable proof which this margin is too small to contain. (Around 1637)

This of course refers to his famous "Fermat's Last Theorem," which was finally proved by Andrew Wiles in 1996 [132]. Most doubt whether Fermat really had a proof, but it is an intriguing part of the history of this famous problem. If Fermat had a solution, he certainly did not have the brilliant one that Wiles found.

Let's talk about some new results that generalize Fermat's *Little* Theorem to matrices. They do not seem to be well known, but are quite pretty. Perhaps they may have applications to some complexity theory problems.

While Fermat's Last Theorem is very hard to prove, the version for polynomials is much easier to resolve. In particular, one can prove that

$$a(x)^m + b(x)^m = c(x)^m$$

has no solutions in non-constant polynomials, for $m > 2$. There are many proofs of this fact. Often changing a problem from integers to polynomials makes it easier.

Let's now turn to study number theory problems for matrices, not for integers nor for polynomials.

35.1 Fermat's Little Theorem

This states,

Theorem 35.1. *If p is a prime and m an integer, then $m^p \equiv m$ mod p.*

There are many proofs of this beautiful theorem. One is based on the following observation:

$$(x_1 + \cdots + x_m)^p \equiv x_1^p + \cdots + x_m^p \bmod p.$$

The proof of this equation follows from the binomial theorem, and the fact that

$$\binom{p}{k} \equiv 0 \bmod p$$

for p prime and $0 < k < p$. Then, simply set all $x_i = 1$, and that yields,

$$(1 + \cdots + 1)^p \equiv 1 + \cdots + 1 \bmod p,$$

which is $m^p \equiv m \bmod p$.

35.2 Matrices

The brilliant Vladimir Arnold once stated:

> There is a general principle that a stupid man can ask such questions to which one hundred wise men would not be able to answer. In accordance with this principle I shall formulate some problems.

Since Arnold is certainly not stupid, he was joking. Yet there is some truth to his statement: it is notoriously easy to raise questions in number theory that sound plausible, hold for many small cases, and yet are impossible to prove. Fermat's Last Theorem exactly fit this category for over three hundred years. Many others come to mind: for example, the Goldbach Conjecture and the Twin Prime Problem.

In any event, Arnold did extensive numerical experiments, to search for a way to generalize Fermat's Little Theorem to matrices. He noticed, immediately, that simply replacing the integer m by a matrix A will not work. For example, consider a matrix A that is nilpotent, but is not zero. Recall a matrix is *nilpotent* provided some power of the matrix is 0. Then, for p large enough, $A^p = 0$ and so clearly $A^p \not\equiv A \bmod p$.

Thus, Arnold was forced to extend the notion of what it meant for a theorem to be "like" Fermat's Little Theorem. After extensive experiments he made make the following conjecture about the trace of matrices. Recall that the $\text{trace}(A)$ is the sum of the diagonal of the matrix A.

Conjecture 35.1. Suppose that p is a prime and A is a square integer matrix. Then, for any natural number $k \geq 1$,

$$\text{trace}(A^{p^k}) \equiv \text{trace}(A^{p^{k-1}}) \bmod p^k.$$

In his paper, published in 2006 [6], he found an algorithm that could check his conjecture for a fixed size $d \times d$ matrix and a fixed prime. He then checked that it was true for many small values of d and p, yet he did not see how to prove the general case.

Finally, the general result was proved by Alexander Zarelua in 2008 and independently by others [138]:

Theorem 35.2. *Suppose that p is a prime and A is a square integer matrix. Then, for any natural number $k \geq 1$,*

$$\mathrm{trace}(A^{p^k}) \equiv \mathrm{trace}(A^{p^{k-1}}) \bmod p^k.$$

An important special case is when $k = 1$, recall Arnold did prove this case,

$$\mathrm{trace}(A^p) \equiv \mathrm{trace}(A) \bmod p.$$

For an integer matrix A, the corresponding coefficients of the characteristic polynomials of A and A^p are congruent $\bmod p$. This, in effect, generalizes the above statement about traces.

35.3 Factoring

Arnold's conjecture—now theorem—has some interesting relationship to the problem of factoring. *If* we assume that factoring is hard, then his theorem can be used to prove lower bounds on the growth of the traces of matrix powers.

Let's start with a simple example. I will then show how we can get much more powerful statements from his theorem. Consider a single integer matrix A, and look at the following simple algorithm, where $n = pq$ is the product of two primes.

1. Compute $\alpha = \mathrm{trace}(A^n) \bmod n$.
2. Then compute the greatest common divisor of $\alpha - k$ and n for all $k = 1, \ldots, \log^c n$ where c is a constant.

Clearly, if this finds a factor of n it is correct. The only question is, when will that happen? Note, by Arnold's theorem,

$$\alpha \equiv \mathrm{trace}(A^{pq}) \bmod p$$
$$\equiv \mathrm{trace}(A^q) \bmod p.$$

Also,

$$\alpha \equiv \mathrm{trace}(A^{pq}) \bmod q$$
$$\equiv \mathrm{trace}(A^p) \bmod q.$$

The key observation is: suppose that the traces of the powers of the matrix A grow slowly. Let $\alpha \bmod p \neq \alpha \bmod q$, and let one of these values be small. For some k, then the value $\alpha - k$ will be zero modulo p and not zero modulo q. Thus, the gcd computation will work.

All this needs is a matrix so that the trace of its powers grow slowly, and yet are different. I believe that this is impossible, but am not positive.

We can weaken the requirements tremendously. For example, replace one matrix A by a family of integer matrices $A_1, \ldots, A_k$. Then define $\alpha(m)$ as

$$\sum_{i=1}^{k} \lambda_i \text{trace}(A_i^m),$$

where all λ_i are integers.

Now the key is the behavior of the function $\alpha(m)$. In order to be able to factor this function must have two properties:

1. There must be many values of m so that $\alpha(m)$ is small;
2. The values of $\alpha(p)$ and $\alpha(q)$ should often be distinct.

If these properties are true, then the above method will be a factoring algorithm. Thus, if factoring is hard, then any $\alpha(m)$ that is non-constant must not have many small values. This can be easily made into quantitative bounds. I know that some of these results are proved, but their proofs often depend on deep properties from algebraic number theory. The beauty of the connection with factoring is that the proofs are simple—given the strong hypothesis that factoring is hard.

Since I am open to the possibility that factoring is easy, as you probably know, I hope that there may be some way to use these ideas to attack factoring. But either way I hope you like the connection between the behavior of matrix powers and factoring.

35.4 Gauss's Congruence

There are further matrix results that also generalize other theorems of number theory. For example, the following is usually called Gauss's congruence:

Theorem 35.3. *For any integer a and natural number m,*

$$\sum_{d\mid m} \mu\left(\frac{m}{d}\right) a^d \equiv 0 \bmod m.$$

If $n = 1$, then $\mu(n) = 1$; if n is divisible by the square of a prime, then $\mu(n) = 0$; if n is divisible by k distinct primes, then $\mu(n) = (-1)^k$.

Here $\mu(n)$ is the famous Möbius function:

$$\mu(n) = \begin{cases} 1, & \text{if } n = 1, \\ 0, & \text{if } n \text{ is divisible by the square of a prime,} \\ (-1)^k, & \text{if } n \text{ is divisible by } k \text{ distinct primes.} \end{cases}$$

By the way, why does Johann Gauss get everything named after him? We should create a complexity class that is called GC, for "Gauss's Class." Any suggestions?

Alexander Zarelua [138] proves that Gauss's congruence generalizes nicely to matrices:

Theorem 35.4. *For any integer matrix A and natural number m,*

$$\sum_{d\mid m} \mu\left(\frac{m}{d}\right) \operatorname{trace}(A^d) \equiv 0 \bmod m.$$

As an example, let $m = pq$ the product of two primes. Then this theorem shows that we can determine

$$\alpha = \operatorname{trace}(A^p) + \operatorname{trace}(A^q) \bmod pq$$

for any integer matrix A. What intrigues me is, if α is less than pq, then we get the *exact* value of

$$\operatorname{trace}(A^p) + \operatorname{trace}(A^q).$$

Can we do something with this ability? In particular, can it be used to get *some* information about the values of p and q?

35.5 Open Problems

I believe there are two interesting types of questions. The first is what can we say about the growth of such sums of matrix traces? Can we improve on known bounds or give shorter proofs based on the hardest of factoring?

Second, a natural idea is to look at other Diophantine problems and ask what happens if the variables range over matrices? Which known theorems remain true, which become false, and which become open?

35.6 Notes

This appeared as fermats-little-theorem-for-matrices in the blog. I thank Qiaochu Yuan for some corrections.

Part IV
On Mathematics

Chapter 36
A Curious Algorithm

Zeke Zalcstein worked, before he retired a few years ago, on the boundary between mathematics and computational complexity. He started his career in mathematics, and then Zeke moved into computational complexity. His Ph.D. advisor was John Rhodes, who is an expert on many things, including the structure of semigroups; Rhodes has recently worked on the $P=NP$ question himself.

I have known Zeke since I was a graduate student at Carnegie-Mellon and he was faculty there. Later, Zeke worked at a number of universities, including Stony Brook, and spend the latter part of his career as a program director at CISE/NSF.

Zeke loves to laugh, he is funny, he knows tons of mathematics—especially algebra; yet he is a —character in the best sense of the word. For example, he is very particular about what he eats and drinks, even his water cannot be just any water. When Zeke was at Stony Brook he discovered there was a water source in a certain open field with the best water in the world. Not just Stony-Brook, the world. The water source was a faucet coming out of the ground, in an otherwise empty field. I asked Zeke where the water came from, the faucet was simply connected to a pipe that came out of the ground, no label, no sign, nothing. Did it come from a spring, from a well, or from the city water main? Zeke said he just knew it was great water.

Unfortunately, one day Zeke fell and broke his arm so it was impossible for him to drive for several weeks. While Zeke was incapacitated a certain graduate student was kind enough to make periodic trips out to the faucet, fill up plastic jugs with the special water, and deliver them to Zeke.

I once was telling this story to John Hennessy, who is now the President of Stanford University. He started to laugh, and confirmed the story: John had been the graduate student who was kind enough to help Zeke.

Enough, on to the main topic. Zeke and I worked together on a number of problems over the years, and I will talk about one with a curious history, a neat proof, and an interesting application that never really happened.

36.1 The Problem and Our Results

The problem is: what is the space complexity of the word problem for the free group on two letters? This is not how we first heard the problem, but it is equivalent to what we were asked. The free group on two letters "a" and "b" is the group where the only relationships are:
$$aa^{-1} = a^{-1}a = bb^{-1} = b^{-1}b = 1.$$

It is called the free group since these are the minimal relations that can hold in any group. As usual the *word problem* is to determine given any word over a, a^{-1}, b, b^{-1} whether or not it equals 1. For the rest of the chapter, when we say word problem, we mean "the word problem for the free group on two letters."

Zeke and I proved [101]:

Theorem 36.1. *The word problem is in* L.

Actually we proved more:

Theorem 36.2. *A probabilistic log-space machine with a one-way read only input tape can solve the word problem with error at most* ε, *for any* $\varepsilon > 0$.

There is a simple linear time algorithm for the word problem. The algorithm uses a pushdown store, which is initially empty. The algorithm processes each input symbol x as follows: If the top of the pushdown is y and $xy = 1$, then pop off the top of the pushdown; if not, then push x onto the pushdown. Then, go to the next input symbol. When there are no more input symbols accept only if the pushdown is empty. The algorithm clearly runs in linear time, and it is not hard to show it is correct.

This algorithm uses linear space: a string that starts with many a's will, for example, require the pushdown to hold many symbols. Thus, the goal is to find a different algorithm that avoids using so much space.

In the next two sections I will explain the "curious history" of the question: who I "think" first asked the question, and why they may have asked the question. Then, I will explain how we solved the problem and proved our theorem. You can skip the next two sections and get right to the proof method. No history. No background. But, I hope that you find the history and motivation interesting. Your choice.

36.2 Whose's Question is It?

My memory for technical conversations is usually pretty good, but this question on the space complexity of the word problem has a murky history. My version is that at a STOC conference, years ago, Juris Hartmanis mentioned a related problem to a number of us over a coffee break. I instantly liked the problem, and during the next few months Zeke and I worked on the problem, until we found a solution.

At the next conference, FOCS, we told Juris our solution to "his" problem. Juris said that he liked the problem, liked our solution, but he had never asked us the problem. Who knows. At least we did not call our paper, "On a Conjecture of Hart-manis." I still believe, sometimes, perhaps Juris told it to us, but I must be confused. Anyway I think you will like the method we used to solve it.

36.3 Why Ask It?

The problem *someone* asked us, was not what is the space complexity of the word problem for the free group on two letters. Instead we were asked a more "language" type question that is the same as the word problem. We were asked, what is the space complexity of a certain language *D2*?

Before defining *D2* it may help to explain where *D2* comes from. Define the following language *D* over the alphabet that contains two types of parentheses: "[" and "]" and "(" and ")". The language *D* contains the empty string, and if *w* is in *D* then so are

$$[w] \text{ and } (w).$$

Thus,

$$[\ [\ (\ ) \ ] \ ]$$

is in the language, but

$$[\ (\ [\ ) \ ] \ ]$$

is not.

Note, that *D* is the language of expressions with two types of parentheses that are *well nested*; sometimes it is called the Dyck language. Each [must have a matching], and also each (must have a matching). Moreover, the two types of parentheses must not get in each others way, thus

$$[\ (\ ] \ )$$

is not in *D*. The Dyck language *D* is a context-free language, and is central to the theory of context-free language theory.

The following is an easy theorem:

Theorem 36.3. *The language D is in* L.

In order to prove this theorem we need a definition: call a string [-*balanced* provided it has the same number of occurrences of [as]; call a string (-*balanced* provided it has the same number of occurrences of (as). Clearly, for any *w* in *D*, it must be both [-balanced and (-balanced. However, this alone is not enough, since the string

$$[\ (\ ] \ )$$

is both [-balanced and (-balanced, but is not in *D*. However, the next lemma characterizes *D* in terms of the notions of balanced.

Lemma 36.1. *Suppose the string w satisfies the following property: if $w = x[y]z$ and y is [-balanced, then y must also be (-balanced. Also the dual: if $w = x(y)z$ and y is (-balanced, then y must also be [-balanced. Then, w is in D.*

The proof of this is a simple induction on the definition of D. Note, it is easy for a L machine to check membership in D, since these properties all dependent on the ability of a log-space machine to be able to count and keep track of a fixed number of positions in the input string.

The language $D2$ is a extension of the Dyck language that allows a different type of pairing of the symbols. This is the question someone asked us—I guess not Hartmanis. Sometimes $D2$ is called a "two-sided" Dyck language. Again the empty string is in $D2$. If w is in $D2$, then so are

$$[w] \text{ and }]w[$$

and also so are

$$(w) \text{ and })w($$

The point of $D2$ is that the parentheses have a type but no "direction."

The difficulty with $D2$ is there does not seem to be a counterpart to the balancing lemma. A string of the form

$$\ldots [w] \ldots$$

could have the [match with a] that comes earlier. This is why I do not know a direct counting method to recognize $D2$, and is probably why someone asked us the question. I wish I could remember who.

36.4 Our Solution

Zeke and I knew two insights, one trivial and the other based on a classic result from group theory. The first is that $D2$ is really the same as accepting the words from a free group on two letters.

$$[\to a \text{ and }] \to a^{-1}$$

and

$$(\to b \text{ and }) \to b^{-1}.$$

The second insight is that the free group on two letters has a faithful representation over 2×2 integer valued matrices.

Theorem 36.4. *There are matrices A and B, such that the mapping $a \to A$ and $b \to B$ is a faithful representation of the free group on a, b.*

Let me explain what this means. Consider the following two matrices:

$$A = \begin{bmatrix} 1 & 2 \\ 0 & 1 \end{bmatrix}$$

and

$$B = \begin{bmatrix} 1 & 0 \\ 2 & 1 \end{bmatrix}.$$

Each matrix is invertible, and further their inverses are also integer matrices The map that sends

$$a \to A \text{ and } a^{-1} \to A^{-1} \text{ and } b \to B \text{ and } b^{-1} \to B^{-1}$$

defines an isomorphism. That means we can replace the word problem by: does a sequence of matrices over A, B, A^{-1}, B^{-1} equal the identity matrix I. This transformation of the word problem into a problem about matrices is the key to our results. For example,

$$aba^{-1}b \to ABA^{-1}B.$$

Here is how Zeke and I used this transformation to prove the our theorem. We show how to check whether or not

$$M = M_1 \times M_2 \times \cdots \times M_n$$

is equal to I where each M_i is from A, B, A^{-1}, B^{-1}. The obvious way to do this is to compute the matrix M and see if it is equal to I. The difficulty with this approach is that the matrix M may have entries that are too large and cannot be stored in log-space.

We solve this with the Chinese Remainder Theorem. Suppose that p is a prime with at most $O(\log n)$ bits. Let $M_p = M \bmod p$. A log-space machine can pass over the input and compute the product M_p: this means that we do all the arithmetic modulo p, but we still are multiplying 2×2 matrices. Then, the machine checks whether or not $M_p = I$. If it does not, then clearly $M \neq I$, and we can reject the input. The machine does this for all primes p of the given size. If all $M_p \equiv I \bmod p$, then the machine accepts.

We claim that this algorithm is correct. The insight is that if M is not equal to I, then $M - I$ has some non-zero entry, which cannot be too large. Then, by the Chinese Remainder Theorem we have a contradiction, and so the algorithm is correct. This uses the well known fact,

$$\prod_{p \leq t} p \geq c^t$$

where the product is over primes and c is a constant greater than 1.

Finally, the probabilistic result follows in essentially the same way. The only difference is that now the machine randomly selects one prime p to check. We then argue the non-zero entry of $M - I$ is unlikely to be divisible by a randomly selected prime.

36.5 Karp-Rabin

What does the Karp-Rabin pattern matching algorithm [78] have to do with our result on the word problem? Indeed. Dick Karp and Michael Rabin are both famous for so many things, but one of my favorites among their results is their randomized pattern matching algorithm. Rabin told me that the way he first thought of this algorithm is that he used our insight about the free group to get the hashing function they needed for the algorithm. It is not too hard to see that the same transformation modulo a prime makes for a good hashing function.

Unfortunately for Zeke and for me they quickly got rid of the matrix ideas and replaced them by a much simpler method. But Michael has repeatedly told me that the matrix ideas played a role in his initial thinking.

36.6 Open Problems

We were able to prove that even more than the word problem for free groups is in L. Suppose that G is any linear group over the rationals. Then, the word problem for this group can also be solved in L. Thus, many infinite groups have word problems that can be done in log-space.

The power of matrix representation theory is something that you may be able to use to solve other problems. I believe that we have not made as much use of the power of representation methods in other parts of computer science.

Now if I could only recall who initially asked me the problem ...

36.7 Notes

This appeared as the-word-problem-for-free-groups in the blog. Alex Nikolov said he "loves the Karp-Rabin" algorithm, and I completely agree. Jeffrey Shallit raised an open problem about the freeness of certain matrices that seems very hard to resolve.

Chapter 37
Edit Distance

Richard Bellman created the method of dynamic programming in 1953. He was honored with many awards for this and other seminal work, during his lifetime. We probably all know dynamic programming, use it often to design algorithms, and perhaps take it for granted. But, it is one of the great discoveries that changed the theory and practice of computation. One of the surprises–at least to me–is given its wide range of application, it often yields the fastest known algorithm.

I never had the honor of meeting Bellman, but he wrote a terrific short auto-biographical piece that contained many interesting stories [15]. One story that I recall was how, when he was at the Rand Corporation, he often made extra money playing poker with his doctor friends in the evenings. He usually won, which convinced them that he was using "deep mathematics" against them–perhaps probability theory. Bellman said the truth was much simpler: during the day the doctors were stressed making life and death decisions, so at night when they played poker they "let go." During the day Bellman could just look out at the Pacific Ocean from his office and think about problems. At night he played tight simple poker, and he cleaned up.

One other story from this piece made a long-lasting impression on me. When he was famous he would often be invited to places to give talks, like many scientists. However, he *always* insisted that he must be picked up at the airport by a driver of a *white limo*. I understand the driver part, I understand the limo part, but insisting on the color was a very nice touch.

37.1 The Edit Distance Problem

The edit distance problem (EDP) is, given two strings

$$a = a_1, a_2, \ldots, a_n \text{ and } b = b_1, b_2, \ldots, b_m$$

determine their *edit distance*. That is, what are the *fewest* operations that are needed to transform string a into b. The operations allowed are: insert a character, delete a character, and substitute one character for another. In general their costs can be arbitrary, but for us we will consider only the basic case where insert/delete have unit cost and substitute has cost two. This makes sense, since substitute is the same as one delete and one insert.

This distance is sometimes called the Levenshtein distance, named after Vladimir Levenshtein, who defined it in 1965. The notation of edit distance arises in a multitude of places and contexts, I believe that the notion has been repeatedly discovered. It also has many generalizations where more complex operations are allowed.

People really care about EDP. There are very good heuristic algorithms that are used, for example, in the biological community to solve EDP. One is called BLAST. It is implemented in many languages and runs on everything; there are even special purpose hardware devices that run BLAST. Clearly, there is need for solutions to EDP. Unlike theory algorithms, BLAST has no provable error bounds; unlike theory algorithms, BLAST seems, in practice, to be good enough for the scientists who use it. However, I have no doubt that they would prefer to get the optimal answer—the computational cost of getting the optimal answer cannot be too great.

37.2 Upper and Lower Bounds

The dynamic programming algorithm for EDP has been discovered and rediscovered many times. I believe that Robert Wagner and Mike Fischer did it first [131], but I will not try to be the perfect historian. Their algorithm runs in time $O(n^2)$ and uses the same space. It is convenient to assume that $n = m$ for the rest of this chapter. With a little bit of work, it is easy to improve the space bound to $O(n)$.

There are papers that prove a lower bound of $\Omega(n^2)$ for EDP. These papers assume a restricted model of computation: only input symbols can be compared for equality. Like any restricted lower bound they give you an idea of what not to do: if you wish to beat the the lower bound, then you must do more than compare symbols for equality. Use the value of the symbols, use bit operations, use randomness, do something other than just compare symbols.

Actually, as I recall *before* the lower bound papers, there was a better upper bound paper that gave an algorithm for EDP that took time $O(n^2/\log n)$. The method used to prove this sub-quadratic algorithm was based on a general idea called *The Four Russians Method*. I never liked the name, I do like the method. As Dan Gusfield points out [62], the four were not even all Russian: perhaps a consequence of the cold war.

The Four Russians Method was first, I believe, used to compute boolean matrix products faster than cubic time. The method has been used since then to solve many other problems.

Still here. Okay an overview of the method is that it is based on trading space for time. A typical Four Russians Method algorithm operates in two phases. In the first phase, it examines the input and cleverly pre-computes values to a large number of subproblems, and stores these values in a look-up table. In the second phase, it uses the pre-computed values from the table to make macro steps during the rest of the computation. The result is usually a reduction in time by a logarithmic factor, while the space becomes as large as the time. The method is not generally practical; however, it is a useful method to know. Ryan Williams has made some progress in a recent paper related to this method [133].

37.3 Bit Complexity

Actually there is a logarithmic factor that is hidden in the quadratic dynamic programming algorithm for EDP, since the algorithm must use $\log n$ size numbers. Thus the bit complexity $O(n^2 \log n)$. There is a clever way to remove the factor of $\log n$ that is due to Dan Lopresti [96]:

Theorem 37.1. *There is an algorithm for EDP that uses $O(n^2)$ bit operations and $O(n)$ space.*

Proof. The key idea is to use only a constant number of bits rather than $\log n$ bits and still run the standard dynamic program.

A disclosure: Lopresti was once a graduate student of mine, and I worked with him on this result.

37.4 Approximation

Legend has it that when faced with unraveling the famous Gordian Knot, Alexander the Great simply cut the knot with his sword. We do this all the time in theory: when faced by a "knotty" problem we often change the problem. The edit distance problem is any not different: years ago the problem of exact distance was changed to approximate edit distance. The goal is to get a fast algorithm that finds the distance to within a small relative error. For example, in the STOC 2009 conference there was a paper "Approximating Edit Distance in Near-Linear Time" by Alexandr Andoni and Krzysztof Onak. They show that there is an algorithm (AO) that in time $n^{1+o(1)}$ gets a relative error of

$$2^{O(\sqrt{\log n} \log \log n)}.$$

This is a huge improvement over the previous work. I will not go into any details; see their paper for the full proofs [5].

This is very pretty work. I have looked at their paper, and I believe that it has many interesting ideas, that may even help solve other problems. My issue is that we still are no closer to solving the edit distance problem. The AO algorithm still makes too large relative error to be practical–what exactly is the factor for reasonable size n? See Chp. 27 on the use of " O-notation."

On the other hand, I am intrigued with their breakthrough. I think that it may be possible to use their new algorithm as a basis of an exact algorithm. The key idea I have in mine is to try to use their algorithm combined with some kind of "amplification trick."

37.5 Amplification and EDP

All of the papers on approximations, to the edit distance problem use "relative" error. Note, if anyone could prove an additive error result that would be very powerful. Suppose that $A(x,y)$ is an algorithm for the EDP. Say that $A(x,y)$ has *additive error* $E(n)$ provided, for all strings x,y of length n,

$$A(x,y) = d(x,y) + O(E(n)).$$

There is a simple amplification trick:

Theorem 37.2. *Suppose there is an algorithm $A(x,y)$ for the EDP that runs in time $T(n)$ and has additive error $E(n)$. Then, for any $k = k(n)$, there is an algorithm $A^*(x,y)$ for the EDP that runs in time $T(cnk)$ for a fixed constant $c > 0$ and has an additive error of $O(E(cnk)/k)$.*

I assume that this is well known, but I thought it would be useful to point out why relative error is the favorite: additive error is too hard. As a simple example, if there is an algorithm that runs in near linear time, i.e. time $n^{o(1)}$, for EDP and has additive error of $n^{1+o(1)}$, then there is be a near linear time exact algorithm for EDP.

I have started thinking about using the AO algorithm and combining it with a different amplification method—see Chp. 40. Perhaps this could lead, finally, to a fast exact algorithm for EDP. Perhaps.

37.6 Open Problems

Thus, the main open problem is clear: find an EDP algorithm that runs in $O(n)$ time. Or one that runs in $O(n \ln n)$ time, or near linear time. Or one that runs in … I see no reason, even after decades of failure, to believe that there could not be an exact algorithm that runs fast and solves the EDP. This is–in my opinion–an important and exciting open problem. Let's finally solve the problem.

There are other, perhaps, more approachable open problems. In real applications the inputs are not worst case, nor are they random. For instance, the BLAST heuristic assumes implicitly a kind of randomness to the inputs. Are there reasonable models of the kinds of inputs that are actually encountered? And for such a model, can we get provable fast algorithms for EDP?

37.7 Notes

This appeared as bellman-dynamic-programming-and-edit-distance in the blog. Nick Black and others refer to Bellman's *Eye of the Hurricane: An Autobiography* as a source of insight into many things—including what it was like being a graduate student at Princeton years ago.

Chapter 38
Protocols

Ashok Chandra is a famous theorist who has done seminal work in the many parts of the foundations of complexity theory. His early work was in program schemas, an area that was central to theory in the 1970's. Chandra then moved into other areas of theory, and among other great things co-invented the notion of Turing Machine Alternation—see Chp. 12.

Later in his career, while still an active researcher, he began to manage research groups: first a small one at IBM Yorktown, then a much larger one at IBM Almaden, and later groups at other major companies including Microsoft. As a manager, one example of his leadership was the invention by his group at Almaden of that little trackpoint device that plays the role of a mouse. The trackpointer eventually appeared on the keyboards of IBM's Thinkpad laptops—now called Lenovo laptops— a wonderful example of successful technology transfer.

Let's talk about mutli-party protocols, an area that Chandra helped create in the 1980's. Then, I will connect this work with a new paper that appeared in FOCS 2009.

I still remember vividly meeting Ashok for the first time: it was at the Seattle airport in 1974 at a taxi stand. We were both on our way to the sixth STOC conference—it was my first STOC. I knew of Ashok, since I had read his beautiful papers on program schemata theory. Somehow we figured out that we were both going to the same hotel and we shared a cab. Our conversation in the cab made me feel I was welcome as a new member of the theory community: I have never forgotten his kindness.

At that STOC Chandra presented some of his recent work on program schemas, "Degrees of Translatability and Canonical Forms in Program Schemas: Part I." Also presented at the conference were these results, among others:

- Leslie Valiant: *The Decidability of Equivalence for Deterministic Finite-Turn Pushdown Automata.*
- John Gill: *Computational Complexity of Probabilistic Turing Machines.*
- Vaughan Pratt, Michael Rabin, Larry Stockmeyer: *A Characterization of the Power of Vector Machines.*

- Stephen Cook, Robert Reckhow: *On the Lengths of Proofs in the Propositional Calculus.*
- Robert Endre Tarjan: *Testing Graph Connectivity.*
- Allan Borodin, Stephen Cook: *On the Number of Additions to Compute Specific Polynomials.*
- David Dobkin, Richard Lipton: *On Some Generalizations of Binary Search.*

The topics then were very different from today: then, there was no quantum theory, no game theory, no learning theory, and many complexity classes had yet to be invented. Yet many of the topics are still being studied today.

Let's now turn to multi-party protocols.

38.1 Multi-Party Protocols

Chandra, Merrick Furst, and I discovered the basic idea of multi-party protocols because we guessed wrong. We were working on proving lower bounds on bounded width branching programs, when we saw that a lower bound on multi-party protocols would imply a non-linear lower bound on the length of such programs.

Let me back up, one summer Merrick Furst and I were invited to spend some time at IBM Yorktown by Ashok. The three of us started to try to prove a lower bound on the size of bounded width programs.

A *bounded width program* operates as follows—see Chp. 19. Suppose that the input is $x_1, \ldots, x_n$. I like to think of a bounded width program as a series of "boxes." Each one takes a state from the left, reads an input x_i which is the same for the box each time, and passes a state to the right. The first box is special and gets a start state from the left; the last box is special too and passes the final state to the right. The total number of states is fixed independent of the length of the computation. Note, also while the boxes read the same input each time, an input can be read by different boxes. The length, the number of boxes, is the size of the bounded width program.

The reason this model is called bounded width is that the number of states is fixed independent of the number of inputs. The power of the model comes from the ability to read inputs more than once. Note, without that ability the model could not compute even some very simple functions.

We guessed that there should be simple problems that required length that is super-linear in the number of inputs, n. This guess was right. What we got completely wrong, like most others, was that we thought the length might even be super-polynomial for some simple functions. David Barrington's famous theorem [14] proved that this was false, see Chp. 19:

Theorem 38.1. *For any boolean function in* NC^1*, the length of a bounded width program for that function is polynomial.*

The connection between bounded width programs and protocols is simple. Suppose that there are $2n$ boxes, which we wish to show is too few to compute some function. Divide the boxes into three groups: L, M, and R. The boxes in L are the first $2n/3$ boxes, the boxes in M are the next $2n/3$ boxes, and L are the last $2n/3$. Think of three players L, M, and R. They each do not have enough bits to determine the function if it depends on all the bits. Yet any two players together have enough bits, possibly, to determine the function: this follows because,

$$2n/3 + 2n/3 > n.$$

Thus, we needed to somehow argue that they needed to move more than a fixed number of bits from one to another to compute the given function. This was the hard part.

IBM Yorktown is in the middle of nowhere. Well it is in the middle of a beautiful area, but then there were no lunch places nearby. So every day we ate in the IBM cafeteria, which was a fine place to eat. However, one day Merrick and I decided to take off and have lunch somewhere else. We had a long lunch "hour." A really long hour. When we got back to IBM Ashok was a bit upset: where have we been? Luckily a theory talk was just about to start and we all went to hear the talk.

Talks are a great place to think. There is no interruption, which was especially true back then since there were no wireless devices. I used the time to try to put together all the pieces that we had discovered. Suddenly I saw a new idea that might be the step we were missing.

As soon as the talk was over I outlined the idea that I had to Ashok and Merrick. They were excited, and to my relief all was forgotten about the long lunch. Working that afternoon we soon had a complete outline of the proof. There were, as usual, lots of details to work out, but the proof would eventually all work. I still think we made progress on this result because of the long lunch and the quiet time of the talk. The result eventually appeared in STOC 1983 [29].

We were excited that we had a lower bound, but we missed two boats. Missing one is pretty bad, but two was tough. The first was our protocol lower bound: this was later vastly improved by much better technology, and the second was that we missed thinking about upper bounds. I do not know if we would have found Barrington's beautiful result; however, we never even thought about upper bounds. Oh well.

38.2 Open Problems

One of the lessons we should have learned, but did not is: do not only work on one direction of an open problem. We got a nice lower bound, but missed the great result of Barrington. I doubt we would have found it, but we never even thought about an upper bound. Never.

38.3 Notes

This appeared as multi-party-protocols-and-focs-2009 in the blog. Stefan Ciobac asked was the the motivation for the bounded width model based on a practical need. I answered the motivation was that it modeled nicely a theoretical model of computation.

Chapter 39
Erdős and the Quantum Method

Paul Erdős needs no introduction to our community, but I will give him one anyway. He won the AMS Cole Prize in 1951 and the Wolf Prize in 1983. He is famous for so many things: an elementary proof of the Prime Number Theorem, the solver of countless open problems, the creator of countless more open problems, and his ability to see to the core of a problem. There are, by the way, several versions of how he and Atle Selberg found their respective elementary proofs of the Prime Number Theorem, and more importantly where the credit should go and to whom. See [59] for a history of the controversy.

Our friends at Wikipedia have a story about Erdős's use of amphetamines, but the version I heard from Ron Graham is a bit different. Ron bet Erdős $500 dollars that he could not stop taking the drug for a month. Exactly one month later Erdős met Ron, took his $500 dollars in winnings, and immediately popped a pill. Erdős said to Ron, "I felt mortal."

Perhaps my favorite story about Erdős is the "dinner conversation story." Apparently Erdős was at a conference and found himself seated at a table of experts all from some arcane area of mathematics–let's call it area A. Paul knew nothing about their area. Nothing. The rest of the table was having a lively discussion about the big open question in their area A. Erdős, of course, could not follow any of the discussion, since he knew none of the terms from A. Slowly, by asking lots of questions, "what does that term mean?" and "why is that true?" and so on, Erdős began to understand the open problem of area A. Once he finally understood the problem, once the definitions and basic facts had been made clear to him, he was able to solve their open problem. Right there. Right then.

During the rest of the dinner Paul explained his solution to the table of experts–who I am sure had mixed feelings. Whether true or not, I think this story captures Erdős's ability to cut to the core of an open problem, and more often than not the core was a combinatorial problem. When it was combinatorial, there was a good chance that he could solve it.

A personal story—alas my Erdős number is only 2—is about the time I told Erdős about the "Planar Separator Theorem." This is of course joint work with Bob Tarjan. After listening politely to me nervously explain the theorem, I was quite junior at the time, Paul smiled and said, "that is not to be expected," I took this as a compliment. It made my day.

39.1 The Probabilistic Method

Erdős and the Quantum Method? What am I talking about? Actually Erdős is famous, of course, for the *probabilistic method*. I believe similar methods had been used by others—again I am not trying to be a perfect historian—but Erdős certainly deserves credit for making the probabilistic method a standard tool that everyone working in theory must know. The method is amazingly powerful—clearly many of today's results would be unprovable without the method.

You probably, no pun intended, know the general method: If you wish to construct an object with some property, often a "random" object can be proved to have the property with positive probability. If this is the case, then there must *exist* some object with the desired property. The best current explanation of this method is the great book of Noga Alon and Joel Spencer [4].

However, the first book on the probabilistic method was the thin small blue book of Erdős and Spencer [46]. The book consisted of a series of chapters, each giving another example of the probabilistic method. There were no motivations, no exercises, no overview, no frills. Just chapter after chapter of "We will prove X. Consider a random object ..." They used this chilling phrase in the book: "it follows by elementary calculation that this formula is true." That always scared me. Often such statements could be checked in a hour or two, while I remember one that took me a few days. The "elementary calculation," in this case, used Stirling's approximation for $n!$ out to four terms. I loved the book.

I still remember my first use of the probabilistic method to solve an open problem. A year before studying the book I was asked about a certain open problem by Arnie Rosenberg of IBM Yorktown. I never got anywhere on it, and eventually forgot about it. After studying the book, one day I ran into Andy Yao at a coffee shop near MIT, and asked him what he was working on. He stated the problem, and it sounded familiar. I suddenly realized two things: one it was the problem Arnie had asked me a year earlier, and second that I knew how to solve it. The answer was the probabilistic method. In about two days Andy and I had worked out all the details and we had solved the problem [98].

39.2 The Quantum Method

I have my doubts about the future of Quantum Computing. But I always think of:

Clarke's First Law. When a distinguished but elderly scientist states that something is possible, he is almost certainly right. When he states that something is impossible, he is very probably wrong. Arthur C(harles) Clarke

So my doubts are probably wrong. However, whether or not quantum computers are ever built is not important once you realize that there is a new proof method based on quantum theory. There is a small but growing collection of theorems that have nothing to do with quantum computation, yet are proved by quantum arguments. This is extremely important. This–to me–changes everything. If quantum theory becomes a new technology for proving theorems, then we have to become experts in it or be left behind.

There is an uncanny parallel with the probabilistic method. The amazing part of the probabilistic method is that it can solve problems that have nothing to do with probability theory. The statement of the theorems do not talk about random objects. The only place randomness is used is *inside* the proof itself. This is the magic of the method. Take a pure statement that is completely deterministic, and prove by probability theory that the statement is true. That is where the power of the probabilistic method lies, in my opinion.

This is beginning to happen with quantum theory. There are now a small number of theorems that can be proved by using quantum arguments. The statements of the theorems have nothing to do with anything quantum. Nothing. I believe that such quantum arguments could parallel the probabilistic method, and help solve many currently open problems. It is early and we do not yet know if this new method can be as powerful as the probabilistic method, but time will tell.

The quantum method yields proofs that look something like this:

- Assume, by way of contradiction, that some classic result CR is true.
- Use simulation, or another method, to show that this implies that a quantum result QR is true.
- Reach a contradiction by showing that QR is too good and violates some known result from quantum theory.

Thus, the classic result CR is false. Very cool.

39.3 Examples of The Quantum Method

One of the major players in this new area is Ronald de Wolf. He has been able to prove a number of theorems using the quantum method. I will highlight one of his results on locally decodable codes. His other results include: a result on the degrees of symmetric functions, and another on a lower bound on matrix rigidity. The key in all of these is that the results have nothing in their *statements* about quantum theory. The degree of the symmetric function is a classic measure, and matrix rigidity is a classic notion. And so on. Again, notice the parallel with the probabilistic method.

The quantum method is in its infancy, as de Wolf has pointed out to me. There are a few examples, yet there is no underlying structure, hence still no sense where the method will go. Will we look back, in a few years, and see a few isolated results, or is this the beginning of a new chapter in theory? Only time will tell, but as an outsider I am excited by the initial results. Others are working on results that may also fall under this umbrella: apparently Scott Aaronson is one of them. Of course I am not surprised, since Scott is one of the experts on all things quantum.

Let's turn to one important example of the quantum method due to Lordanis Kerenidis and de Wolf [80]. They prove a lower bound on the size of certain codes. An error-correcting code is said to be a *locally decodable code* (LDC) if a randomized algorithm can recover any single bit of a message by reading only a small number of symbols of a possibly corrupted encoding of the message with an adaptive decoding algorithm. A (q, δ, ε)-locally decodable code encodes n-bit strings x into m-bit code words $C(x)$. For any index i, the bit x_i can be recovered with probability $1/2 + \varepsilon$ with q queries even if δm bits have been corrupted. The theorem of Kerenidis and de Wolf is:

Theorem 39.1. *For all δ, ε there exists a $\lambda > 0$ such that for all n every $(2, \delta, \varepsilon)$-locally decodable code has length $m \geq 2^{\lambda n}$.*

Here is a outline of their proof reproduced here with de Wolf's kind permission. Given a 2-query LDC $C : \{0,1\}^n \to \{0,1\}^m$ we will proceed as follows to get a lower bound on m.

Let's assume that the classical 2-query decoder recovers the bit x_i by choosing a random element (j, k) from some perfect matching M_i of $[m]$. Then queries bits j and k of the (possibly corrupted) codeword, and returns their XOR as its guess for x_i. It was already known [79] from Jonathan Katz and Luca Trevisan that this "normal form" can be achieved, at the expense of reducing the success probability by a small amount.

Now consider the quantum state ("quantum code for x") $|\phi_x\rangle$ equal to,

$$1/\sqrt{m} \sum_{j=1}^{m} (-1)^{C(x)_j} |j\rangle$$

This is the uniform superposition over all entries of the codeword $C(x)$ (with bits turned into signs). This state is an m-dimensional vector, so it has $\log(m)$ qubits. There is a quantum measurement (slightly technical but not hard) that, when applied to this state, gives us the XOR of the two bits of a random element from M_i, for each i of our choice, and hence allows us to make the same prediction as the classical 2-query decoder.

Thus we have a quantum state that allows us to predict each of the n bits of x. It follows from known quantum information theory (the random access code lower bound of Ashwin Nayak [111]) that the number of qubits is $\Omega(n)$. Accordingly, we have $\log(m) = \Omega(n)$, hence $m \geq 2^{\Omega(n)}$.

The exponential lower bound matches–up to the small constant factors in the exponent—to the Hadamard code, which is a 2-query LDC. This is still the only super-polynomial lower bound known for LDC's.

39.4 Open Problems

The obvious open questions are, can we use the quantum method to solve other open problems? Or to get cleaner proofs of old theorems? Or to extend and sharpen known theorems? I do not understand the limits of this new method, but here are some potential problems that could be attacked with the quantum method:

- Circuit lower bounds.
- Data structure lower bounds.
- Coding theory lower bounds.
- Separation theorems, i.e. P$\neq$NP.
- Other suggestions?

One more point: de Wolf suggested that " very metaphorically, quantum physics stands to complex numbers as classical physics stands to real numbers." I pointed out to him that there is a famous quote due to Jacques Hadamard that is relevant: *The shortest path between two truths in the real domain passes through the complex domain.*

Finally, I would like to thank de Wolf for his generous help with this chapter. As usual all opinions, errors, and mistakes, are mine.

39.5 Notes

This appeared as erdos-and-the-quantum-method in the blog. There were many positive comments on the potential of the quantum method by Pascal Koiran, Michael Nielsen, and John Sidles. It was also noticed that Andy Drucker and Ronald de Wolf have a new survey out of the quantum method, in which they cite this post as partial motivation.

Chapter 40
Amplifiers

Andy Yao is a Turing Award winner, a long time friend, a co-author, and one of the great theorists of all time. He has solved many open problems, and helped create many important new areas of complexity theory.

I want to talk about one of Andy's famous results, the XOR Lemma, while placing it in a larger context. Throughout science one of the recurrent themes is the power of methods that are able to amplify something: the "thing" can be a physical object or a signal, or can be a mathematical property. In all cases, amplification methods often lead to great progress and create new insights.

Andy spent many years, as I did, at the computer science department at Princeton. One of his great traits was the ability to focus on one problem, usually until he solved it. Many of us try to think in parallel about many things at once, but in my opinion, Andy is a serial processor. A very powerful serial processor.

I once went, years ago, to see Andy about a new result of mine: the fact that the permanent has a random self-reduction. I told him the result and added that I felt it could be important. He listened politely and checked his calendar, and told me that he would be available at a certain time in ten days hence. That's the way he operated: focused and steady.

You could also guess that this was his style just from looking at his desk. Typically it either had nothing on it or perhaps at most one sheet of paper. One sheet. Some desks in the department looked like a bomb had recently gone off and scattered papers everywhere; others were less disordered. But Andy's was special.

One December the graduate students came to me to ask about trying to create some fun event for the upcoming holiday party. I came up with the "whose-desk-is-it-contest?" They took an aerial photograph of each faculty member's desk. At the party each guest was given a sheet and their task was to match the picture of the desk with the faculty member. It was a silly game, that we all seemed to enjoy, and besides the winner got a fun prize. I believe out of almost one hundred entries no one missed matching Andy's desk to Andy.

Let's turn now to talk about the main topic on amplifiers in the spirit of the XOR Lemma for Yao.

40.1 Physical Amplifiers

There are countless examples of *amplification* techniques. Here are some that work
with physical systems:

• **The lever.** This is the familiar system that can be used to amplify a force. The
famous Archimedes of Syracuse once said, "Give me a lever long enough and a
fulcrum on which to place it, and I shall move the world."

• **The microscope.** The first microscope was apparently made in the late 1500's
in the Netherlands. Hans Lippershey, and Sacharias Jansen are usually credited with
the invention. One repeating theme in all these amplifier inventions is that it is often
hard to be sure who invented them first. This is true of older amplifiers, like the
microscope, and as you will see of more modern ones too. I am not sure why this is
the case; perhaps someone can shed some light on why?

• **The telescope.** Like the microscope, the telescope seems to have come from
the Netherlands also in the early part of the 1600's. The great Galileo Galilei did
not invent the telescope, but improved it, and also was perhaps the first to really
understand it value. The inventors are usually listed as Hans Lippershey, Sacharias
Jansen, and Jacob Metius.

• **The relay.** Joseph Henry invented the first relay in 1835. You all probably
know well how a basic relay works: a small current flows through a coil that moves
an arm. The arm when moved allows another circuit to be completed and a larger
current can then flow.

• **The vacuum tube.** Lee De Forest is credited with creating the basic vacuum
tube, in 1907. He had a grid that could be charged negatively or positively, and in
this way control a current that flowed across the tube. In his version there was not a
true vacuum in the tube; that improvement was made later by GE scientists.

• **The transistor.** This is perhaps the most important of all amplifiers, without
which the entire base of modern technology would be impossible. Try building an
iPhone with vacuum tubes. It was invented on 17 November 1947 by John Bardeen
and Walter Brattain, at AT&T Bell Labs. Quickly, William Shockley saw the vast
potential and is often viewed as the father of the transistor.

40.2 Physical and Algorithmic Amplifiers

• **The PCR method.** Kary Mullis won the Nobel Prize in 1993 for his discovery
of this method that allows the amplification of even one strand of DNA. It is an
amazing result for several reasons. Of course without PCR modern TV shows like
CSI would be impossible. Even more important, PCR is one of the central tools in
all of bio-technology. See Chp. 41 for more details.

There is some controversy over who invented what when. I have no insider information, but I have an opinion. What makes PCR so different for the biologists is
that PCR needed no new biology. From a theory point of view what Mullis did was
use existing ideas to create a new "algorithm." PCR uses basic facts about DNA that

were long known: the double strands break apart when heated; they tend to pair up A-T and G-C when cooled; there is an extension process that uses an enzyme called DNA polymerase. It was discovered by Arthur Kornberg who won the Nobel Prize in 1953.

Mullis' contribution, I believe, was putting together the pieces to form the PCR algorithm. An analogy for us is any famous algorithm. Typically, such an algorithm only uses the operations that any computer can perform, but their order is critical. And moreover understanding their effect is what makes the algorithm new and exciting.

40.3 Algorithmic Amplifiers

• **The XOR lemma.** This is the famous result of Yao that shows how to amplify the hardness of a boolean function. Roughly his result shows that if f is a boolean function that is weakly predictable, then the new function defined by

$$F(x_1, \ldots, x_t) = f(x_1) \oplus \cdots \oplus f(x_t)$$

is almost completely unpredictable, provided t is large enough. Like the other amplifiers this one also has an interesting history. The first published proofs were by others—no doubt Andy had a proof, but the details were done later on.

• **The parallel repetition method.** This is the famous result of Ran Raz on showing that winning n games in parallel decreases exponentially in n [116]. This result also has an interesting history. Is there something about amplifiers that makes for complex history?

The result was first stated by Lance Fortnow and Mike Sipser, who thought that playing in parallel was the same as playing serially [50]. Quickly, it was realized that this was not obvious, and the search began for a proof. After many partial results, including some work I did with Anne Condon and Jin-Yi Cai on a special case [26], Ran resolved the whole problem. His proof is quite hard and there are now other proofs that are easier to follow.

• **The matrix block trick.** This is a folk idea—I believe—that can be used in various contexts to amplify the effect of some theorem about matrices. Here is a simple example: suppose that A is an $n \times n$ matrix whose permanent you ish to approximate. For $n = 3$, create the matrix $B = A^{\otimes n}$, which consists of n copies of A as diagonal blocks,

$$B = \begin{pmatrix} A & 0 & 0 \\ 0 & A & 0 \\ 0 & 0 & A \end{pmatrix}$$

Then it is easy to see that the value of the permanent of B is the value of the permanent of A to the nth power. This shows that if there is a way to approximately compute the value for B, then there is a way to get a much better approximation to the value of A. This trick can be used for many other matrix valued problems.

• **The repetition method.** This is perhaps the most basic amplification method. In order to increase something, just do the same thing many times. This works for error correcting codes, for example. Sending the message many times will increase the probability of getting all the bits of a message. One of the problems is that it is not very efficient, but it is very useful precisely because it relies on no structure of the problem.

Another nice example of the method is to amplify the correctness probability of a BPP computation. If the computation has a probability of $2/3$ of being correct, then t repeats of the computation with *independent* coin flips will have an error probability that is at most $1/3^t$. One of the major thrusts in this area, of course, is to do better than this simple repetition method.

• **The tensor power trick.** This is a trick that is explained beautifully in Terence Tao's great blog [128]. Let me try to hopefully explain what is up. The trick—he calls it that—is to amplify an inequality and eventually get a better one. He uses the following transformation: take a function $f : X \to \mathbb{C}$ and replace it by

$$f^{\otimes n}(x_1, \ldots, x_n) = f(x_1) \cdots f(x_n).$$

He shows that using this transformation can often improve an inequality and make it sharper. In his hands this simple amplification idea can work wonders.

• **The direct product method.** I have to stop somewhere, so I will stop with a recent paper by Russell Impagliazzo, Valentine Kabanets and Avi Wigderson on "Direct Product Testing" from STOC 2009 [71]. This is really another type of repetition amplification, similar to the last idea of Tao's. The *direct product code* of a function f gives its values on all possible k-tuples $((x_1), \ldots, (x_k))$—the direct product code consists of k-tuples $(f(x_1), \ldots, f(x_k))$. This is used to amplify hardness in many constructions. They prove some beautiful new results on how to test how close a set of k-tuples is to a direct product.

40.4 Open Problems

One simple open problem is to create a list of methods that fall under this umbrella. I know I have just listed some of the amplification methods.

The other problem is to create new methods, that are currently missing. Here are a couple of potential ones:

1. Can we make a PCR-like amplifier for other types of molecules? In particular what about an amplifier for proteins? I do not believe that one is known. There are very indirect methods to make more of a protein, but that is not an amplifier. If I am right there is a Nobel lurking here—but proteins have none of the neat properties of DNA so it probably is a very tough problem.
2. One the top open problems today in computational game theory is finding good approximations to various game/economic problems. One approach, perhaps, is to try and create an amplifier here. Suppose that G is a bi-matrix game. Can we

construct another game H so that an approximate strategy for H yields an even better strategy for G? One must be a bit careful since there are hardness results, so the amplifier must not be in conflict with these results.

3. Another of my favorite problems is the edit distance problem. There are many very good approximation methods for this famous problem. Can we build an amplifier here that could help improve these results? More precisely, given two strings x and y can we construct new strings x' and y' so that a good approximation to the edit distance between x' and y' yields an even better one between x and y?

40.5 Notes

This appeared as the-role-of-amplifiers-in-science in the blog. Pat Morin pointed out Chernoff's bound is another example of an amplifier.

Chapter 41
Amplifying on the PCR Amplifier

Kary Mullis is not a theorist, but is a chemist who won the 1993 Nobel Prize for inventing the PCR method—which is an amplifier for DNA. He is also an alumnus of Georgia Tech, where I am these days. Mullis is famous, perhaps our most famous graduate, but he does have a quirky personality. Perhaps to create a brilliant idea like PCR took such a thinker.

I want to discuss the details of his PCR method, which I hope you will enjoy. It really is an *algorithm*. Really. I think this is, at least partially, why there was such a controversy swirling around whether or not Mullis did *enough* to win a Nobel Prize. He certainly is one of the few who have been honored with a Nobel Prize for the invention of an algorithm; for example, Tjalling Koopmans and Leonid Kantorovich won their economics Nobel Prizes for linear programming in 1975. But there are few others. Also, PCR raises a pure theory problem that I hope someone will be able to resolve. More on that shortly.

Mullis invented PCR while he was employed at Cetus, now a defunct company. He got a $10,000 invention award, while Cetus eventually sold the patent rights for 300 million dollars. This was back when this was real money; even today that is a pretty good amount for a single patent. Mullis was not happy with the size of his award, which is not hard to understand.

This reminds me of a story about IBM. They have an award for the *best invention of the year*, which comes along with a nice, but relatively modest, cash award. Years ago the inventor of a certain encoding trick for improving the density of disk storage units was awarded this prize. Then he was awarded the same prize the following year, for the same invention. This is like being "Rookie of the Year" for two years in a row. Since the invention made IBM billions of dollars in additional sales of disk units, they needed some way to thank him. So he got the best invention of the year, for the same invention, twice. Perhaps Cetus could have done something like this too.

Back to PCR. There was a huge court battle over whether the Cetus patent for PCR was valid or not. Several famous biologists, including some Nobel Laureates, testified against Cetus and Mullis, arguing that the PCR method was already known, that Mullis did not invent anything, and therefore that the patent should be thrown out. Mullis won.

I have read parts of the trial transcript and other accounts, and I think I have a theory why there was such outcry against Mullis. First, he is—as I stated already—a bit quirky. If you have any doubts read his book [109]—"Dancing Naked in the Mind Field." Second, there was a great deal of money and prestige at stake. The patent, if valid, was extremely valuable. Finally, there was—and this is my two cents—the algorithm issue. Mullis invented no new chemistry, he discovered no new biological process, he found nothing new in the usual sense. What he did that was worthy of a Nobel was "just" put together known facts in a new way.

To me this is like arguing that the famous RSA algorithm is not new: we all knew about modular arithmetic, about primes, about Fermat's Little Theorem, and so on. So how could Ron Rivest, Adi Shamir, and Len Adleman be said to have invented something? They did. As theorists, we appreciate the huge breakthrough they made—but the biologists do not usually think in terms of algorithms, so they had trouble with PCR.

Let's now turn to the Polymerase Chain Reaction (PCR) algorithm and the open question about it that I wish to raise. If you want you can jump to next to the last section where I discuss the open question, but I worked hard on the next sections— the DNA diagrams were tricky—so at least look at them.

41.1 DNA

The cool thing about PCR is that you do not need to know a lot about the deep structure of DNA in order to understand the method. There are a couple of key facts that you do, however, need to know. A single strand of DNA can, for us, be viewed as a string over a four-letter alphabet $\{A,T,G,C\}$, where each letter stands for one of four nucleic acids. Thus,

$$A \to T \to G \to C \to T$$

is an example of a single strand of DNA. The bonds denoted by the $\to$ symbol are strong bonds. It is important that the bonds are directional.

Two single pieces of DNA can form a double strand of DNA, which consists of two such sequences placed together: one running left to right, the other right to left. Where the letters meet they follow the two famous Watson-Crick rules:

$$A - T \text{ and } G - C.$$

The $-$ symbol denotes a weak bond.

Thus, the following is an example of a double strand of DNA:

$$A \rightarrow T \rightarrow G$$
$$|\quad |\quad |$$
$$T \leftarrow A \leftarrow C$$

Note, the top strand goes in the opposite direction of the bottom strand.

Mullis used a few simple operations that can be performed on DNA. One is that DNA can be *heated*. When a double strand of DNA is heated properly (not too hot, and not too fast) it breaks nicely into the top and bottom strands. Thus, the above double strand would break into:

$$A \rightarrow T \rightarrow G \ \text{ and } \ C \rightarrow A \rightarrow T.$$

A key point is that

$$C \rightarrow A \rightarrow T \ \text{ and } \ T \leftarrow A \leftarrow C$$

are the same.

Another simple operation is that DNA can be *cooled*. In this case single strands will try to find their Watson-Crick mates and bond. Note, the above two strings could re-form again into:

$$A \rightarrow T \rightarrow G$$
$$|\quad |\quad |$$
$$T \leftarrow A \leftarrow C$$

The key is *could*. In order for this to happen they must find each other. If they were alone in a test tube, it would be very unlikely to happen. But, if there are many copies of each type, then many of the above double strands would form.

The last operation that Mullis uses is the most interesting, and it relies on the enzyme *polymerase*. Consider the following piece of DNA made of two single strands:

$$A \rightarrow T \rightarrow G \rightarrow G \rightarrow G$$
$$|\quad |$$
$$C \leftarrow C$$

Note that the bottom is incomplete. How do we get the remaining part of the bottom strand constructed? The answer is to add the enzyme polymerase and a sufficient supply of each nucleic acid A, T, G, C. Then, polymerase moves along and adds the missing bases—shown in bold—one at a time. In this case first a C, then an A, and finally a T.

$$A \rightarrow T \rightarrow G \rightarrow G \rightarrow G$$
$$|\quad |\quad |\quad |\quad |$$
$$\mathbf{T} \leftarrow \mathbf{A} \leftarrow \mathbf{C} \leftarrow C \leftarrow C$$

An important point is that polymerase only completes the bottom if it is already started. Thus a single strand of DNA will not be changed at all. The top must have some of the bottom started already; actually the bottom part must be at least a few bases. One more key point to make is that polymerase only adds in the direction of the DNA strand. Thus, in this case nothing will happen:

$$A \rightarrow T \rightarrow G \rightarrow G \rightarrow G$$
$$| \quad |$$
$$T \leftarrow A$$

All these steps—heating, cooling, and polymerase—were known for decades before Mullis's work; amazingly that is all he needs to do the PCR algorithm.

41.2 Basic PCR

I would like to explain what I will call the basic method first, then I will explain the full method. Suppose that you start with a double strand of DNA of the form:

$$\alpha \rightarrow \ldots \rightarrow \beta$$
$$| \quad \ldots \quad |$$
$$\hat{\alpha} \leftarrow \ldots \leftarrow \hat{\beta}$$

where α, β, $\hat{\alpha}$ and $\hat{\beta}$ are chains formed of the nucleic acids A,T,G,C. Also, α and β have enough length for polymerase to work. Put the following all together in a test tube. The double strands of the above DNA that you wish to copy; lots of single copies of A,T,G,C; lots of polymerase. Then add one more ingredient: lots of copies of the *primers*, which are single strands of DNA of the form α and $\hat{\beta}$.

Then all you do is heat, cool, heat, cool, for some number of cycles. The magic is that this is enough to make an exponential number of copies of the original DNA. To see this consider what will happen after the first heat-and-cool cycle. The above will become:

$$\alpha \rightarrow \ldots \rightarrow \beta$$
$$|$$
$$\hat{\beta}$$

and

$$\alpha$$
$$|$$
$$\hat{\alpha} \leftarrow \ldots \leftarrow \hat{\beta}$$

This follows since during heating the double strand breaks into the top and bottom strands; then during cooling the most likely event is for the abundant primers to attach to the top and the bottom. It is very unlikely for the double strands to reattach together—since there are so many primers. Next, polymerase does its magic and after a while the above becomes:

$$\alpha \rightarrow \ldots \rightarrow \beta$$
$$| \quad \ldots \quad |$$
$$\hat{\alpha} \leftarrow \ldots \leftarrow \hat{\beta}$$

and

$$\alpha \to \ldots \to \beta$$
$$| \quad \ldots \quad |$$
$$\hat{\alpha} \leftarrow \ldots \leftarrow \hat{\beta}$$

Thus, we have doubled the number of strands of the DNA—we have an amplifier.

41.3 Full PCR

The full PCR method can copy a selected part of a long strand of DNA. This is really the method that Mullis invented. The ability to copy only part of a much longer strand is of huge importance to biologists. Curiously, it is exactly the same method as the basic one. The only difference is the analysis of its correctness is a bit more subtle: I find this interesting, since this happens all the time in the analysis of algorithms.

Suppose that you start with the following double strand of DNA:

$$x \to \alpha \to \ldots \to \beta \to y$$
$$| \quad | \quad \ldots \quad | \quad |$$
$$\hat{x} \leftarrow \hat{\alpha} \leftarrow \ldots \leftarrow \hat{\beta} \leftarrow \hat{y}$$

As before, put the following all together in a test tube. The double strands of the above DNA; lots of single copies of A, T, G, C; lots of polymerase. Then, add one more ingredient: lots of copies of the *primers*, which are single strands of DNA of the form α and $\hat{\beta}$. Then it is just a bit harder to see that the following becomes amplified:

$$\alpha \to \ldots \to \beta$$
$$| \quad \ldots \quad |$$
$$\hat{\alpha} \leftarrow \ldots \leftarrow \hat{\beta}$$

Thus, only the DNA *between the primers* is exponentially increased in number. Other, partial pieces of DNA will grow in quantity, but only at a polynomial rate. This is the PCR algorithm.

41.4 Can PCR Amplify One Strand of DNA?

Can PCR copy one piece of DNA? I have asked this over the years of my biology friends, and have been told that they believe that it can. However, usually PCR is used to amplify many strands of DNA into a huge amount of DNA. So the problem that I want to raise today is: can theory help determine the ability of PCR to amplify even one piece of DNA? The obvious lab protocol would be to prepare test tubes that have exactly one piece and see how the PCR method works. The trouble with this is simple: how do we make a test tube that is guaranteed to have exactly one piece of the required DNA? I believe that this is not completely trivial—especially for short strands of DNA.

41.5 The Amplifier Sensitivity Problem

Note, even if biologists can prepare test tubes with one strand, the following problem is interesting since it could have applications to other amplification methods. The problem is an ideal version of trying to decide whether PCR can amplify one strand or not, *without the ability to create test tubes with one strand.*

I have thought about this issue some, and would like to present it as a pure theory problem. I will give a simple version, but you can probably imagine how we could make the problem more realistic.

Imagine that you have boxes—think test tubes. A box can have a number of balls—think DNA strands—inside: the number can range from 0 to a very large number. You can hold a box, but no amount of shaking or any other trick will tell you anything about its contents. If you hold it up to the light, you see just the box.

However, you do have a *detector*, that given a box does the following: if the box has fewer than k balls in it, then the detector says *none*; if the box has more than k balls in it, the detector says *yes there are balls*. Once a box is given to the detector, the box is destroyed.

Your task is to determine the value of k. This value is always at least 1, but is otherwise, secret. So far there is not much you can do. I will let you have a number of operations that you can use in your quest to find the value of k:

1. You can ask for a new box. This will be filled with a varying number of balls, but the number of balls b in the box will always be vastly larger than k.
2. You can ask that the contents of a box be randomly divided into l boxes. This operation destroys the original box and returns to you l boxes. The balls of the original are distributed in a uniformly random manner into the l boxes. For example, you might wish to split the contents of a box into two new boxes. If the original contained b balls, then the new ones will have x and y where $x + y = b$, and the partition is random.

That is all you can do. The question is: can you devise a protocol that will get a good estimate on the secret value of k? A related question: suppose that I tell you that k is 1 or 2. Can you devise a protocol that determines which is true? Note, you also have access to randomness so the protocol is assumed to be random, and therefore the results need only hold with high probability.

41.6 Open Problems

The main question is to solve the amplifier sensitivity problem and find k. Can we develop a protocol that can solve this? Also, can we discover a protocol that tells $k = 1$ from $k = 2$?

Note, even if PCR has been shown in the lab to be able to amplify one piece of DNA, these protocols may be of use in the future with other types of amplifying mechanisms. As science moves down into the depths of nano scale, I can certainly imagine other applications of a protocol that can determine how sensitive an amplifier is.

Please solve these problems. Or prove that they cannot be solved.

41.7 Notes

This appeared in amplifying-on-the-pcr-amplifier in the blog. There some discussion of the box problem, and it seems to still be open.

Chapter 42
Mathematical Embarrassments

Terence Tao is one of the greatest mathematicians of our time. Tao has already solved many long-standing open problems, which earned him a Fields Medal. What I cannot understand about him is how he can be so productive. He is the author of one of the best blogs in the world, he writes original articles that solve hard open problems, he writes great survey articles on a wide range of topics, and he is the author of terrific books on topics from additive combinatorics to partial differential equations. He is impressive.

I want to talk about a topic that he has discussed: *mathematical embarrassments*. The name is mine, but I hope he will agree with spirit of this term.

There is an old story about Godfrey Hardy and John Littlewood, who, published many great papers together. Steven Krantz tells a version of the story in his fun book, Mathematical Apocrypha [85]:

> It is said that Landau thought "Littlewood" was a pseudonym for Hardy so that it would not seem like he wrote all the papers.

I think Tao might consider using this trick and inventing a co-author or two. It would help make the rest of us feel better.

Let's now turn to mathematical embarrassments.

42.1 Mathematical Embarrassments

A mathematical embarrassment (ME) is a problem that should have been solved by now. An ME usually is easy to state, seems approachable, and yet resists all attempts at attack. There may be many reasons that they are yet to be solved, but they "feel" like they should have been solved already.

42.2 Some Examples

• **The π and e problem.** The problem is to prove that $\pi + e$ and $\pi - e$ are both transcendental numbers. We know that one of these must be transcendental. For if both were algebraic, then so would

$$\frac{1}{2}(\pi + e + \pi - e) = \pi,$$

which contradicts the known fact that π is transcendental.

Similarly, at least one of $\pi + e$ and πe must be transcendental, for otherwise

$$x^2 - (\pi + e)x + \pi e$$

would be an algebraic polynomial with transcendental roots. It seems ridiculous that we are able to show the transcendence of π and e, but not these numbers.

• **The linear recursion zero problem.** An ME stated by Tao is a problem about linear recurrences [128]. Given a linear recurrence over the integers,

$$a_n = c_1 a_{n-1} + \cdots + c_d a_{n-d}$$

with initial conditions, does there exist an k so that $a_k = 0$? He only asks for a decision procedure; he says that it is

> ... faintly outrageous that this problem is still open; it is saying that we do not know how to decide the halting problem even for "linear" automata!

• **Mersenne composites.** The problem is to prove that for an infinite number of primes p,

$$2^p - 1$$

is composite. It is an open question whether there are infinite number of Mersenne primes. Currently, only 47 Mersenne primes are known. It seems more likely that there are an infinite number of Mersenne composites. Leonhard Euler has shown that

Theorem 42.1. *If $k > 1$ and $p = 4k + 3$ is prime, then $2p + 1$ is prime if and only if $2^p \equiv 1 \bmod 2p + 1$.*

Thus, if $p = 4k + 3$ and $2p + 1$ are both primes, then $2^p - 1$ is a composite number. So an approach to the problem is to show there are infinitely many p's of the above form.

• **The Jacobian conjecture.** This is one of my favorite problems outside theory proper. Consider two polynomials $f(x,y)$ and $g(x,y)$. The Jacobian Conjecture asks when is the mapping:

$$(x,y) \rightarrow (f(x,y), g(x,y))$$

one to one? The values of x, y are allowed to be any complex numbers. Clearly, a simple necessary condition is that the mapping be locally one to one: this requires that the determinant of the Jacobian of the mapping to be everywhere non-zero. That is, there must exist a nonzero constant c such that for all x and y

$$\det \begin{pmatrix} f_x(x,y) & g_x(x,y) \\ f_y(x,y) & g_y(x,y) \end{pmatrix} = c.$$

If the value of c were not a constant, it is a non-trivial polynomial, which must take the value zero for some values x', y' over $\mathbb{C}$. So for those values, the mapping will not even be one to one locally.

Note, $f_x(x,y)$ is the value of the partial derivative of f with respect to x, and $f_y(x,y)$ is the value of the partial derivative of f with respect to y. The Jacobian Conjecture is that this necessary condition is sufficient.

This is the conjecture for two dimensions; there is a similar one for higher dimensions, but this case is already open. Note, the maps can be messy; for example, the mapping

$$(x,y) \to (x+y^7, y)$$

is one to one. If you compose this with some linear invertible transformation, then the mapping can start to look very strange.

The Jacobian Conjecture is related to another famous theorem called the Ax—Grothendieck theorem [11].

Theorem 42.2. *Suppose that $f : \mathbb{C}^n \to \mathbb{C}^n$ is a polynomial function. Then, if f is one-to-one, then f is onto.*

The proof is based on the following trivial observation: if a map from a finite set to itself is one-to-one, then it is onto. Amazing. The full proof is based on a simple compactness argument and this simple observation.

• **The existence of explicit non-linear lower bounds.** One of the great ME for theory is that we have no circuit lower bounds that are non-linear on any explicit problem. The natural proof and other barriers do not seem to be relevant here. I do not believe that they stop a lower bound of the form:

$$n \log^* n.$$

• **The BPP hierarchy problem.** The problem is to show that $\mathsf{BPTIME}(n^{100})$ is more powerful than $\mathsf{BPTIME}(n)$. Even with respect to oracles, we do not know if this is true—see Chp. 14. It is hard to believe that such a basic question is still open. There is a partial result due to Jin-Yi Cai, Ajay Nerurkar, and D. Sivakumar [27].

• **The number of planar graphs.** How many labeled planar graphs are there, with a given number of vertices n? If $P(n)$ is the number of such graphs, then the best known bounds are of the form:

$$n! c_1^n \leq P(n) \leq n! c_2^n$$

where $c_1 < c_2$. How can we not know the number of planar graphs?

Well it turns out that I am embarrassed—Omer Giménez and Marc Noy *know*. They have proved—in 2008—an asymptotic estimate for the number of labeled planar graphs:

$$g \times n^{-7/2} \gamma^n n!$$

where g and γ are explicit computable constants. This a beautiful result, which solves a long standing ME [55].

There is more known about 2-connected planar graphs—see the paper by Edward Bender, Zhicheng Gao, and Nicholas Wormald for some pretty results [16]. One that I like very much is that they can show that a random 2-connected planar graph is highly likely to contain many copies of any fixed planar graph:

Theorem 42.3. *For any fixed planar graph H, there exist positive constants c and δ such that the probability that a random labeled 2-connected planar graph G with n vertices has fewer than cn vertex disjoint copies of H is $O(e^{-\delta n})$.*

• **The Barrington problem.** David Barrington's famous theorem shows that polynomial size bounded width computations can compute all of NC^1. He does this by using computations over simple groups. An open question is can simple groups be replaced by solvable groups? I think the conventional wisdom is that this must be impossible. But there seems to be no progress on this problem. The reason it is interesting is that the construction of Barrington fails easily for solvable groups, but that is not a proof.

• **The $n^{d/2}$ problem.** This problem is given a series of integers

$$x_1, \ldots, x_n$$

is there a subset of size d that sums to 0? How can order $n^{d/2}$ be the best possible running time?

This algorithm is essentially the same as the exponential algorithm for the knapsack problem that I talked about in Chp. 20. Recall, that the knapsack problem is to find a 0-1 vector x so that

$$a_1 x_1 + \cdots + a_n x_n = b.$$

The method re-writes the problem as

$$a_1 x_1 + \cdots + a_m x_m = b - (a_{m+1} x_{m+1} + \cdots + a_n x_n)$$

where $m = n/2$—we can assume that n is even. Then, all the left hand sums are computed, and all the right hand sums are computed too. There is a solution to the original problem if and only if these sets have a value in common. Since there are $2^{n/2}$ of them, this method takes time $2^{n/2}$ times a polynomial in n.

• **The Hilbert subspace problem.** Given a Hilbert space and a linear operator A, is there a subspace S so that it is non-trivial, invariant under A, and closed? Invariant under A means that for each x in S, Ax is also in S. This is a classic result for finite Hilbert spaces, but is has long been open in general. Several special cases of this problem have been resolved, but not the general case.

42.3 Open Problems

What are your favorite ME?

42.4 Notes

This appeared as mathematical-embarrassments in the blog. There were some additional examples of their favorite ME's. Some wondered why something was an ME and not a MD (mathematical disease). Thanks again to Neil Dickson, Gil Kalai, Michael Mitzenmacher, András Salamon, John Sidles, Terence Tao, Wei Yu, and many others.

Chapter 43
Mathematical Diseases

Underwood Dudley is a number theorist, who is perhaps best known for his popular books on mathematics. The most famous one is *A Budget of Trisections*, which studies the many failed attempts at the ancient problem of trisecting an angle with only a ruler and a compass [44]. This problem is impossible, yet that has not stopped some people from working day and night looking for a solution. Trying to find such a solution is an obsession for some; it's almost like they have a malady that forces them to work on the problem.

I plan on talking about other mathematical obsessions. They are like diseases that affect some, and make them feel they *have* to work on certain mathematical problems. Perhaps P=NP is one?

Dudley's book is quite funny, in my opinion, although it does border on being a little bit unkind. As the title suggests, in "A Budget of Trisections," he presents one attempt after another at a general method for trisecting any angle. For most attempts he points out the angle constructed is not exactly right. For others he adds a comment like:

> This construction almost worked, if only the points *A* and *B* and *C had* really been co-linear it would have worked. Perhaps the author could move ...

His book is about the kind of mathematical problems that I will discuss today: problems that act almost like a real disease.

I cannot resist a quote from Underwood that attacks bloggers. Note he uses "he" to refer to himself in this quote:

He has done quite a bit of editing in his time–the College Mathematics Journal for five years, the Pi Mu Epsilon Journal for three, the Dolciani Mathematical Expositions book series (six years), and the New Mathematical Library book series (three years). As a result he has a complete grasp of the distinction between "that" and "which" (very rare) and the conviction that no writing, including this, should appear before the public before passing through the hands, eyes, and brain of an editor. **Take that, bloggers!**

(Bold added by me.)
Oh well.

43.1 What Is a Mathematical Disease?

Late 2009 was the flu season in Atlanta, and many I knew were getting it. There is another type of "bug" that affects mathematicians—the attempt to solve certain problems. These problems have been called "diseases," which is a term coined by the great graph theorist Frank Harary. They include many famous problems from graph theory, some from algebra, some from number theory, some from complexity theory, and so on.

The symptoms of the flu are well known—I hope again you stay away from fever, chills, and the aches—but the symptoms for a mathematical disease (MD) are less well established. There are some signs however that a problem is a MD.

1. A problem must be easy to state to be a MD. This is not sufficient, but is required. Thus, the Hodge-Conjecture will never be a disease. I have no clue what it is about.
2. A problem must seem to be accessible, even to an amateur. This is a key requirement. When you first hear the problem your reaction should be: *that is open?* The problem must *seem* to be easy.
3. A problem must also have been repeatedly "solved" to be a true MD. A good MD usually has been "proved" many times—often by the same person. If you see a paper in arXiv.org with many "updates" that's a good sign that the problem is a MD.

Unlike real diseases, MD's have no known cure. Even the solution of the problem will not stop attempts by some to continue working on it. If the proof shows that something is impossible—like the situation with angle trisection—those with the MD will often still work hard on trying to get around the proof. Even when there is a fine proof, those with the disease may continue trying to find a simple proof. For example, Andrew Wiles' proof of Fermat's Last Theorem has not stopped some from trying to find Pierre de Fermat's "truly marvellous proof."

43.2 Some Mathematical Diseases

Here are some of the best known MD's along with a couple of lesser known ones. I would like to hear from you with additional suggestions. As I stated earlier Harary, was probably the first to call certain problems MD's. His original list was restricted to graph problems, however.

• **Graph Isomorphism:** This is the classic question of whether or not there is a polynomial time algorithm that can tell whether two graphs are isomorphic. The problem seems so easy, but it has resisted all attempts so far. I admit to being mildly infected by this MD: in the 1970's I worked on GI for special classes of graphs using a method I called the beacon set method.

There are some beautiful partial results: for example, the work of László Babai, Yu Grigoryev, and David Mount on the case where the graphs have bounded multiplicity of eigenvalues is one of my favorites [13]. Also the solution by Eugene Luks of the bounded degree case is one of the major milestones [102].

I would like to raise one question that I believe is open: Is there a polynomial time algorithm for the GI problem for *expander graphs*? I have asked several people of late and no one seem to know the answer. Perhaps you do.

- **Group Isomorphism:** This problem is not as well known as the GI problem. The question is given two finite groups of size n are they isomorphic? The key is that the groups are presented by their multiplication tables. The best known result is that isomorphism can be done in time $n^{\log n + O(1)}$. This result is due to Zeke Zalcstein and myself and independently Bob Tarjan [100]. It is quite a simple observation based on the fact that groups always have generator sets of cardinality at most $\log n$.

I have been affected with this MD for decades. Like some kind of real diseases I get "bouts" where I think that I have a new idea, and I then work hard on the problem. It seems so easy, but is also like GI—very elusive. I would be personally excited by any improvement over the above bound. Note, the hard case seems to be the problem of deciding isomorphism for p-groups. If you can make progress on such groups, I believe that the general case might yield. In any event p-groups seem to be quite hard.

- **Graph Reconstruction:** This is a famous problem due to Stanislaw Ulam. The conjecture is that the one vertex deleted subgraphs of a graph determine the graph up to isomorphism, provided it has at least 3 vertices. It is one of the best known problems in graph theory, and is one of the original diseases that Harary discussed.

I somehow have been immune to this disease—I have never thought about it at all. The problem does seem to be solvable; how can all the subgraphs not determine a graph? My only thought has been that this problem somehow seems to be related to GI. But, I have no idea why I believe that is true.

- **Jacobian Conjecture:** This is a famous problem about when a polynomial map has an inverse—see Chp. 42. Suppose that we consider the map that sends a pair of complex numbers (x, y) to $(p(x, y), q(x, y))$ where $p(x, y)$ and $q(x, y)$ are both integer polynomials. The conjecture is that the mapping is 1-1 if and only if the mapping is locally 1-1. The reason it is called the Jacobian Conjecture is that the map is locally 1-1 if and only if the determinant of the matrix

$$\begin{pmatrix} p_x(x,y) & q_x(x,y) \\ p_y(x,y) & q_y(x,y) \end{pmatrix}$$

is a non-zero constant. Note, $p_x(x, y)$ is the partial derivative of the polynomial with respect to x. The above matrix is called the Jacobian of the map.

This is a perfect example of a MD. I have worked some on it with one of the experts in the area—we proved a small result about the problem. During the time we started to work together, within a few months the full result was claimed twice. One of the claims was by a faculty member of a well known mathematics department. They even went as far to schedule a series of "special" talks to present the great proof. Another expert in the area had looked at their proof and announced that it was "correct." Eventually, the talks were canceled, since the proof fell apart.

• **Crypto-Systems:** This is the quest to create new public key crypto-systems. While factoring, discrete logarithm, and elliptic curves are the basis of fine public key systems, there is a constant interest in creating new ones that are based on other assumptions.

Some of this work is quite technical, but it seems a bit like an MD to me. There are amateurs and professionals who both seem to always want to create a new system. Many times these systems are broken quite quickly—it is really hard to design a crypto-system.

A recent example of this was the work of Sarah Flannery and David Flannery in creating a new system detailed in their book *In Code* [49]. The book gives the story of her discovery of her system, and its eventual collapse.

• **P=NP:** You all know this problem.

43.3 Open Problems

What are other MD's? What is your favorite? Why do some problems become diseases? While others do not?

I would love to see some progress made on group isomorphism—I guess I have a bad case of this disease. I promise that if you solve it I will stop thinking about it. Really.

43.4 Notes

This appeared as on-mathematical-diseases in the blog. There were many great additions to my list. I had an entire another post on their comments as more-on-mathematical-diseases. Thanks to all including Ted Carroll, Gil Kalai, Shiva Kintali, Chris Peikert, Joseph O'Rourke, Mark Reid, Aaron Sterling, Charles Wells, and others. I especially want to thank Gil for his long and quite interesting list of MD's.

Chapter 44
Mathematical Surprises

Gil Kalai is one of the great combinatorialists in the world, who has proved terrific results in many aspects of mathematics: from geometry, to set systems, to voting systems, to quantum computation, and on. He also writes one of the most interesting blogs in all of mathematics named *Combinatorics and more*; I strongly recommend it to you.

Let's talk about surprises in mathematics. I claim that that there are often surprises in our field—if there were none it would be pretty boring. The field is exciting precisely because there are surprises, guesses that are wrong, proofs that eventually fall apart, and in general enough entropy to make the field exciting.

The geometer Karol Borsuk asked in 1932: Can every convex body in $\mathbb{R}^d$ be cut into $d + 1$ pieces of smaller diameter? This became the Borsuk Conjecture, which was proved in low dimensions and for special classes of convex bodies—for example, Hugo Hadwiger proved in 1946 that it was true for all smooth convex bodies. The intuition seemed to be that the result was going to eventually be proved; it was just a matter of time.

The shock, I think more than a surprise, came when Jeff Kahn and Kalai proved in 1993 that for d large enough the answer was not $d + 1$. On the contrary, they showed that for d large enough, we need *at least*

$$c^{\sqrt{d}}$$

pieces. Notice that this is a lower bound, and does not imply that $c^{\sqrt{d}}$ pieces are enough. Here $c > 1$ is a fixed constant. I think it is fair to say that $c^{\sqrt{d}}$ is really *different* from $d + 1$—no? Their paper is short and brilliant: you definitely should take a look at it [72].

44.1 Surprises In Mathematics

There are many additional surprises in mathematics. I have listed a few that you may find, I hope, interesting. I tried to break them into "types."

44.2 We Guessed The Wrong Way

Sometimes in mathematics and theory, the conventional wisdom is wrong—we believe that X is true, but it is false. The evidence for X could have been wrong for many reasons: it was based on small or special cases, it was based on false analogies, it was based on a view that the world would be simpler if X were true, or it was based on our lack of imagination.

The "we" is used to denote the community of all who work on mathematics and theory: it includes you and me. So this comment, and all later ones, are addressed to me as well as everyone else. I have guessed wrong many times, which has led me to work extremely hard on trying to prove a false statement. Perhaps, everyone else is a better guesser than I am, but I think we all have our limits.

Here are two examples of this:

• **Nondeterministic space closed under complement:** Neil Immerman and Robert Szelepcsényi independently solved the famous LBA problem [69, 126]. They used a very clever counting method to prove that NLOG was closed under complement. This was, I believe, a pretty good example of a surprise. As I discuss in Chp. 18, most had guessed that NLOG would *not* be closed under complement. A great result.

• **A Vision Conjecture:** In the Bulletin of the American Mathematical Society, a conjecture due to Béla Julesz is discussed [42]. He was an expert in human visual perception, and is famous for a conjecture about perception. After many human experiments, over the years, he concluded that if two images had the same first and second order statistics, then a person would find them indistinguishable.

David Freedman and Persi Diaconis founded a simple rule that generates pictures that have the same first, second, and even third order statistics that a random picture has. Yet, their pictures are easily distinguished by a person from a random picture. I quote Persi:

> When we told Julesz, he had a wonderful reaction. "Thank you. For twenty years Bell Labs has paid me to study the Julesz conjecture. Now they will pay me for another twenty years to understand why it is wrong."

44.3 We Accepted a False Proof

Sometimes in mathematics and theory, a proof is given that is false—just wrong. We all are human, we all make mistakes; sometimes those mistakes can be believed for long periods of time.

Here are three examples of this:

• **Four-Color Theorem:** The Four-Color Theorem (4CT) dates back to 1852, when it was first proposed as a conjecture. Francis Guthrie was trying to color the map of counties in England and observed that four colors were enough. Consequently, he proposed the 4CT. In 1879, Alfred Kempe provided a "proof" for the 4CT. A year later, Peter Tait proposed another proof for 4CT. Interestingly both proofs stood for 11 years before they were proved wrong. Percy Heawood disproved Kempe's proof in 1890, and Julius Petersen showed that Tait's proof was wrong a year later.

However, Kempe's and Tait's proofs, or attempts at a proof, were not fully futile. For instance, Heawood noticed that Kempe's proof can be adapted into a correct proof of a "Five-Color Theorem." There were several attempts at proving the 4CT before it was eventually proved in 1976.

• **Hilbert's 16th Problem:** I will not state what Hilbert's 16th problem is—it will not affect this example. The key is that in 1923 Henri Dulac proved a pretty result about the number of limit cycles that a certain family of differential equations could have. He showed that the number was always finite. While the 16th was not his goal, this result helped solve at least part of the Hilbert problem. Dulac's result was considered a great theorem.

Almost sixty years later, Yulij Ilyashenko, in 1981, found a fatal bug in Dulac's proof. Seven years later he and independently Jean Ecalle, Jacques Martinet, Robert Moussu, and Jean Pierre Ramis found correct proofs that showed that Dulac's theorem was correct, even if his proof was not [87].

• **The Burnside Conjecture:** William Burnside is a famous group theorist who proved many wonderful theorems. Perhaps his most famous is the pq theorem that played a major role in starting the quest to classify all finite simple groups:

Theorem 44.1. *A finite group of order $p^a q^b$ where p, q are primes is solvable.*

He also made a number of very influential conjectures. Perhaps the most famous was made in 1902, and became known as the *Burnside Conjecture*: If a group is finitely generated and periodic, then it is finite. Periodic means simply that for any element x in the group, $x^{n(x)} = 1$ for some $n(x)$. This was eventually shown to be false in 1964.

However, a natural question arose immediately, what if the $n(x)$ was the same for all the elements of the group, then would the group be finite? In order to attack this new problem, group theorists broke it up into many cases: $B(m,n)$ is the class of groups that are generated by m elements, where all elements in the group satisfy, $x^n = 1$. Sergei Adjan and Pyotr Novikov proved that $B(m,n)$ is infinite for n odd, $n \geq 4381$ by a long complex combinatorial proof in 1968 [1].

Another group theorist, John Britton, claimed an alternative proof in 1970 [23]. Unfortunately, Adjan later discovered that Britton's proof was wrong.

I once looked at Britton's proof. It is a whole monograph of about 300 pages, with many technical lemmas. Britton needed many constants in his proof, so rather that statements like, "let $v > 191 \times j^2$," he would write "let $v > c_{11} \times j^2$," where c_{11} was a constant to be determined later on. Then, in the appendix he collected all the inequalities that he needed his constants to satisfy, and all he had to do is show that the inequalities were consistent. He did this. Britton had indeed successfully gone through the cumbersome task of checking the consistency of the inequalities, coming up with constants that satisfy all of them simultaneously.

The bug that Adjan discovered was that Britton had made a simple mistake and written down one inequality incorrectly. Unfortunately, fixing the "typo" created a system that was inconsistent, and so the proof was unsalvageable.

44.4 We Misjudged a Problem

Sometimes in mathematics and theory, a problem is believed to be very hard. The problem is believed to be so hard that either no one works seriously on the problem, or people try to create whole new theories for attacking the problem. Then, a elementary proof is given—one that uses no new technology, one that could have been found many years before.

Here are two examples of this:

• **Van der Waerden Conjecture:** Van der Waerden made a conjecture about the permanent in 1926. He conjectured that

$$\text{perm}(A) \geq n! \left(\frac{1}{n}\right)^n$$

for any doubly stochastic matrix A. Further, that equality holds only for the matrix that has all entries equal. A *doubly stochastic* matrix is a non-negative matrix with its rows and columns sums equal to 1.

This seems like a straightforward inequality. Somehow it stood for decades until it was solved independently by Georgy Egorychev and Dmitry Falikman, in 1979 and 1980 [130]. The surprise here is that this "notorious" problem was solved by fairly elementary proofs. They were awarded the prestigious *Fulkerson Prize* for their work, even though the proofs were simple—their result was a breakthrough in the understanding of the permanent.

• **Approximation for Planar TSP:** The classic problem of finding a TSP for planar graphs is well known to be NP-hard. What Sanjeev Arora did in 1996 was to find an approximation algorithm that had eluded everyone for years [7]. His algorithm's running time was nearly-linear and the error $1 + \varepsilon$ for any given $\varepsilon > 0$. Joseph Mitchell found a similar result [108] at almost the same time—I always wonder:

why do problems stay open for years, and then are solved by two researchers independently?

László Lovász told me that these algorithms could have been discovered years before—he thought one of the "amazing" things about them was precisely that they did not use any new tools. They are very clever, and both use important insight(s) into the structure of an optimal TSP tour. Yet, he said, the proof could have been found long before, way before.

We Misjudge an Approach to a Problem

Sometimes in mathematics and theory, someone outlines an approach to solve an open problem, which the rest of the community does not believe is viable. This mistake can lead to an interesting surprise when their method does actually work.

Here are two examples of this:

• **Hilbert's 10th:** The famous Tenth Problem asks for a decision procedure that can tell whether or not a Diophantine equation has an integer solution. Such an equation is of the form:

$$F(x_1,\ldots,x_n) = 0$$

where F is an integer polynomial. The combined work of Martin Davis, Yuri Matiyasevich, Hilary Putnam and Julia Robinson yielded a proof that this problem in undecidable in 1970 [103]. The last step was taken by Matiyasevich.

This result would probably have surprised the great David Hilbert—he likely thought that there was such a procedure. So this result could be in the "we guessed wrong section," but I included it here for a reason.

Julia Robinson showed in 1950 that proving that a certain number theory predicate could be encoded into a Diophantine equation would suffice to solve the entire problem. What Matiyasevich did was to show that this could be done. Before he did this the eminent logician Georg Kreisel had remarked:

Well, that's not the way it's gonna go.

Kreisel's insight was that finding such a predicate would not only prove that the Tenth was undecidable, but would also prove that it was exactly the same as the Halting problem. He thought that perhaps this was too much to expect. He was wrong.

• **Four-Color Theorem (Again):** The 4CT theorem was proven by Kenneth Appel and Wolfgang Haken in 1976. The proof is famous, since it not only solved a long standing conjecture dating from 1852, but made extensive use of computation.

They point out that their result is *simply* carrying out an attack on the problem that was already known; the method is essentially due to George Birkhoff. Others had used the method to get partial results of the form: the four color theorem is true for all planar maps of size $\leq X$. In 1922 Philip Franklin proved 25, and later Walter Stromquist 52—this is not an exhaustive list of all the partial results.

Haken once told me that one reason he felt that they succeeded was that they *believed* that the current attack could be made to work. That there was no need for an entire new approach. Rather hard work and some computer magic could resolve the problem. Perhaps the surprise is that the attack actually worked.

44.5 We Never Thought About That Problem

Sometimes in mathematics and theory, a result is found that is entirely new. Well perhaps not completely new, but is not from a well studied area. The result can be very surprising precisely because not one previously had dreamed that such a result of this type could even be proved.

Here are two examples of this:

• **Collapse Of The Polynomial Time Hierarchy:** Dick Karp and I proved that if SAT has polynomial size circuits, then the polynomial time hierarchy collapses to the second level. The proof is really simple—yet it is one of those basic results about computation [77]. It's one of my own results, so there is the risk that my opinion is biased, but I think most would agree with this assessment.

I always felt that one of key reasons that we found this result is simple: we were probably the first that ever thought about the problem. Previously, all circuits results went the other way: if there was a good algorithm, then there was a good circuit. Indeed, in general having a linear size circuit says nothing about the algorithmic complexity of a problem: a problem can have a linear size circuit for all n and still be undecidable.

One of the things that I always tell my students is this: beware of simple counter-examples. Often there can be a gem hiding inside a trivial counter-example.

• **Voting Theorems:** Kenneth Arrow is a Nobel Prize winner in economics. One of his great results is the *proof* that any reasonable voting system that tries to select among 3 or more outcomes is flawed. The proof of this result is not too hard, but I believe the brilliant insight was the idea that one could prove something so sweeping about voting systems [9]. That seems to me to be a great surprise.

44.6 We Never Connected Those Areas

Sometimes in mathematics and theory, the connection of two disparate areas can be quite surprising.

Here are two examples of this:

• **Prime Numbers and Complex Analysis:** Johann Dirichlet's theorem is one of the first uses of complex analysis to prove a theorem about primes. The theorem states that every arithmetic progression

$$a, a+b, a+2b, a+3b, \ldots$$

contains an infinite number of primes, as long as a and b have no common divisor. Dirichlet introduced certain complex functions that he cleverly connected to the number of primes that were in a given progression. Then he proved his theorem by showing that his functions were non-zero at 1, which proved that there are an infinite number of primes in the progression.

• **Distributed Computing and Topology:** I will end with a connection that I know little about, but still feel that it is safe to say that it is surprising. That is the connection between distributed computing and algebraic topology. Maurice Herlihy and Sergio Rajsbaum have a *primer* that should help explain the connection better than I can [64].

44.7 Open Problems

What is your favorite surprise? What is your favorite type of surprise? Are there other types of surprises? I love to hear your thoughts.

Finally, could there be major surprises in complexity theory in the future? I hope so, I really do.

44.8 Notes

This appeared as surprises-in-mathematics-and-theory in the blog. This had a very lively discussion including comments on other surprises. One of my favorites was John Sidles story of letter John von Neumann wrote a to Norbert Weiner, which analyzed the fundamental physical limits to microscopy. The letter was written in 1946, and in 1973 was shown to fall to the right type of "burning arrows." Thanks to Paul Beame, Arnab Bhattacharyya, Martin Davis, Gil Kalai, Mark Neyer, and others for taking the time to share their thoughts.

ERRATUM

The P=NP Question and Gödel's Lost Letter

Richard J. Lipton

Erratum to: The P=NP Question and Gödel's Lost Letter
DOI 10.1007/978-1-4419-7155-5

Extracts (c.800 words) from pp. 373, 375 & 377 from "Collected Works of Kurt Godel" by Godel, Kurt edited by Feferman, Solomon et al (2003) used by permission of Oxford University Press. Please visit our website: www.oup.com.

The online version of the original article can be found under
DOI 10.1007/978-1-4419-7155-5

Richard J. Lipton
Georgia Institute of Technology
College of Computing
School of Computer Science
Atlantic Drive 801
30332-0280 Atlanta Georgia
USA
rjl@cc.gatech.edu

Appendix A
Gödel Lost Letter

This is the famous lost letter of Kurt Gödel.

Princeton, 20 March 1956

Dear Mr. von Neumann:

With the greatest sorrow I have learned of your illness. The news came to me as quite unexpected. Morgenstern already last summer told me of a bout of weakness you once had, but at that time he thought that this was not of any greater significance. As I hear, in the last months you have undergone a radical treatment and I am happy that this treatment was successful as desired, and that you are now doing better. I hope and wish for you that your condition will soon improve even more and that the newest medical discoveries, if possible, will lead to a complete recovery.

Since you now, as I hear, are feeling stronger, I would like to allow myself to write you about a mathematical problem, of which your opinion would very much interest me: One can obviously easily construct a Turing machine, which for every formula F in first order predicate logic and every natural number n, allows one to decide if there is a proof of F of length n (length = number of symbols). Let $\Psi(F,n)$ be the number of steps the machine requires for this and let $\phi(n) = \max \psi(F,n)$. The question is how fast $\phi(n)$ grows for an optimal machine. One can show that $\phi(n) \geq k \cdot n$. If there really were a machine with $\phi(n) \sim k \cdot n$ (or even $\sim k \cdot n^2$), this would have consequences of the greatest importance. Namely, it would obviously mean that in spite of the undecidability of the Entscheidungsproblem, the mental work of a mathematician concerning Yes-or-No questions could be completely replaced by a machine. After all, one would simply have to choose the natural number n so large that when the machine does not deliver a result, it makes no sense to think more about the problem. Now it seems to me, however, to be completely within the realm of possibility that $\phi(n)$ grows that slowly. Since it seems that $\phi(n) \geq k \cdot n$ is the only estimation which one can obtain by a generalization of the proof of the undecidability of the Entscheidungsproblem and after all $\phi(n) \sim k \cdot n$ (or $\sim k \cdot n^2$) only means that the number of steps as opposed to trial and error can be reduced from N to $\log N$ (or $(log N)^2$). However, such strong reductions appear in other finite problems, for example in the computation of the quadratic residue symbol using repeated application of the law of reciprocity. It would be interesting to know, for instance, the situation concerning the determination of primality of a number and how strongly in general the number of steps in finite combinatorial problems can be reduced with respect to simple exhaustive search.

I do not know if you have heard that "Post's problem", whether there are degrees of unsolvability among problems of the form $(\exists y)\phi(y,x)$, where ϕ is recursive, has been solved in the positive sense by a very young man by the name of Richard Friedberg. The solution is very elegant. Unfortunately, Friedberg does not intend to study mathematics, but rather medicine (apparently under the influence of his father). By the way, what do you think of the attempts to build the foundations of analysis on ramified type theory, which have recently gained momentum? You are probably aware that Paul Lorenzen has pushed ahead with this approach to the theory of Lebesgue measure. However, I believe that in important parts of analysis non-eliminable impredicative proof methods do appear.

I would be very happy to hear something from you personally. Please let me know if there is something that I can do for you. With my best greetings and wishes, as well to your wife,

Sincerely yours,

Kurt Gödel

Kurt Gödel

P.S. I heartily congratulate you on the award that the American government has given to you.

References

1. S. I. Adjan and P. S. Novikov. On infinite periodic groups i, ii, iii. *Izv. Akad. Nauk SSSR. Ser. Mat*, 32:212–244, 1968.
2. Leonard M. Adleman. Two Theorems on Random Polynomial Time. In *FOCS*, pages 75–83, 1978.
3. Eric Allender, Peter Bürgisser, Johan Kjeldgaard-Pedersen, and Peter Bro Miltersen. On the Complexity of Numerical Analysis. In *IEEE Conference on Computational Complexity*, pages 331–339, 2006.
4. Noga Alon and Joel Spencer. *The Probabilistic Method*. (Wiley-Interscience Series in Discrete Mathematics and Optimization, 1992.
5. Alexandr Andoni and Krzysztof Onak. Approximating edit distance in near-linear time. In *STOC '09: Proceedings of the 41st annual ACM symposium on Theory of computing*, pages 199–204. ACM, 2009.
6. V. I. Arnold. On the matricial version of fermat–euler congruences. *Japanese Journal of Mathematics*, 1(1):1–24, 2006.
7. Sanjeev Arora. Polynomial time approximation schemes for euclidean traveling salesman and other geometric problems. *J. ACM*, 45(5):753–782, 1998.
8. Sanjeev Arora and Boaz Barak. *Computational Complexity: A Modern Approach*. Cambridge University Press, 2007.
9. Kenneth Arrow. A difficulty in the concept of social welfare. *Journal of Political Economy*, 58(4):328–346, 1950.
10. E. Artin. The Gamma function. In Michael Rosen, editor, *Exposition by Emil Artin: a selection; History of Mathematics*. American Mathematical Society, 2006.
11. James Ax. The elementary theory of finite fields. *The Annals of Mathematics, 2nd Ser.*, 88(2):239–271, 1968.
12. László Babai, Lance Fortnow, and Carsten Lund. Non-Deterministic Exponential Time Has Two-Prover Interactive Protocols. In *FOCS*, pages 16–25, 1990.
13. László Babai, D. Yu. Grigoryev, and David M. Mount. Isomorphism of Graphs with Bounded Eigenvalue Multiplicity. In *STOC*, pages 310–324, 1982.
14. David A. Mix Barrington. Bounded-Width Polynomial-Size Branching Programs Recognize Exactly Those Languages in NC^1. *J. Comput. Syst. Sci.*, 38(1):150–164, 1989.
15. Richard Bellman. Unpublished.
16. Edward A. Bender, Zhicheng Gao, and Nicholas C. Wormald. The Number of Labeled 2-Connected Planar Graphs. *Electr. J. Comb.*, 9(1), 2002.
17. Charles H. Bennett and Gilles Brassard. An Update on Quantum Cryptography. In *CRYPTO*, pages 475–480, 1984.
18. Lenore Blum, Steve Smale, and Michael Shub. On a Theory of Computation and Complexity over the Real Numbers: NP-completeness, Recursive Functions and Universal Machines. *Bulletin of American Mathematical Society*, 21(1), 1989.
19. Dan Boneh. Twenty Years of Attacks on the RSA Cryptosystem. *Notices of the American Mathematical Society (AMS)*, 46(2):203–213, 1999.
20. Dan Boneh, Richard A. DeMillo, and Richard J. Lipton. On the Importance of Eliminating Errors in Cryptographic Computations. *J. Cryptology*, 14(2):101–119, 2001.
21. Maria Luisa Bonet and Samuel R. Buss. Size-Depth Tradeoffs for Boolean Fomulae. *Inf. Process. Lett.*, 49(3):151–155, 1994.
22. Alfredo Braunstein, Marc Mézard, and Riccardo Zecchina. Survey propagation: an algorithm for satisfiability. *CoRR*, cs.CC/0212002, 2002.
23. John Britton. *The existence of infinite Burnside groups, Word Problems*, pages 67–348. North Holland, 1973.
24. David Brumley and Dan Boneh. Remote timing attacks are practical. *Computer Networks*, 48(5):701–716, 2005.
25. Randal E. Bryant. Graph-Based Algorithms for Boolean Function Manipulation. *IEEE Transactions on Computers*, 35:677–691, 1986.

26. Jin-Yi Cai, Anne Condon, and Richard J. Lipton. PSPACE Is Provable by Two Provers in One Round. *J. Comput. Syst. Sci.*, 48(1):183–193, 1994.

27. Jin-Yi Cai, Ajay Nerurkar, and D. Sivakumar. Hardness and Hierarchy Theorems for Probabilistic Quasi-Polynomial Time. In *STOC*, pages 726–735, 1999.

28. E. Cardoza, Richard J. Lipton, and Albert R. Meyer. Exponential Space Complete Problems for Petri Nets and Commutative Semigroups: Preliminary Report. In *STOC*, pages 50–54, 1976.

29. Ashok K. Chandra, Merrick L. Furst, and Richard J. Lipton. Multi-party protocols. In *STOC '83: Proceedings of the fifteenth annual ACM symposium on Theory of computing*, pages 94–99, New York, NY, USA, 1983. ACM.

30. Ashok K. Chandra, Dexter Kozen, and Larry J. Stockmeyer. Alternation. *J. ACM*, 28(1):114–133, 1981.

31. Zhi-Zhong Chen and Ming-Yang Kao. Reducing Randomness via Irrational Numbers. *CoRR*, cs.DS/9907011, 1999.

32. Edmund M. Clarke. Programming Language Constructs for Which it is Impossible to Obtain "Good" Hoare-Like Axiom Systems. In *POPL*, pages 10–20, 1977.

33. Richard Cleve. Towards Optimal Simulations of Formulas by Bounded-Width Programs. *Computational Complexity*, 1:91–105, 1991.

34. Edith Cohen. Estimating the Size of the Transitive Closure in Linear Time. In *FOCS*, pages 190–200, 1994.

35. Paul Cohen. *Set Theory and the continuum hypothesis*. New York: Benjamin, 1966.

36. Anne Condon. The Complexity of Stochastic Games. *Inf. Comput.*, 96(2):203–224, 1992.

37. Anne Condon and Richard J. Lipton. On the Complexity of Space Bounded Interactive Proofs (Extended Abstract). In *FOCS*, pages 462–467, 1989.

38. Bryon Cook, Andreas Podelski, and Andrey Rybalchenko. Termination proofs for systems code. In *PDLI*, 2006.

39. S. Cook. The Complexity of Theorem-Proving Procedures. In *Proceedings of Third Annual ACM Symposium on Theory of Computing*, New York, NY, 1971.

40. Stephen A. Cook. Soundness and Completeness of an Axiom System for Program Verification. *SIAM J. Comput.*, 7(1):70–90, 1978.

41. Stephen A. Cook and Pierre McKenzie. Problems Complete for Deterministic Logarithmic Space. *J. Algorithms*, 8(3):385–394, 1987.

42. Persi Diaconis and David Freedman. On the statistics of vision: the Julesz conjecture. *Math. Psych.*, 24(2):112–138, 1981.

43. Danny Dolev, Cynthia Dwork, Nicholas Pippenger, and Avi Wigderson. Superconcentrators, Generalizers and Generalized Connectors with Limited Depth (Preliminary Version). In *STOC*, pages 42–51, 1983.

44. Underwood Dudley. *Budget of Trisections*. Scientia Books, 1987.

45. Calvin C. Elgot and Michael O. Rabin. Decidability and Undecidability of Extensions of Second (First) Order Theory of (Generalized) Successor. *J. Symb. Log.*, 31(2):169–181, 1966.

46. Paul Erdős and Joel Spencer. *Probabilistic methods in combinatorics*. New York, Academic Press, 1974.

47. Walter Feit and John Thompson. Solvability of groups of odd order. *Pacific Journal of Mathematics*, 13(1):775–1029, 1963.

48. Philippe Flajolet and Bob Sedgewick. *Analytic Combinatorics*. Cambridge University Press, 2009.

49. Sarah Flannery and David Flannery. *In Code*. Workman Publishing, 2001.

50. Lance Fortnow and Michael Sipser. Retraction of Probabilistic Computation and Linear Time. In *STOC*, page 750, 1997.

51. Steven Fortune and John E. Hopcroft. A Note on Rabin's Nearest-Neighbor Algorithm. *Inf. Process. Lett.*, 8(1):20–23, 1979.

52. Rusins Freivalds. Probabilistic Two-Way Machines. In *MFCS*, pages 33–45, 1981.

53. M. R. Garey, Ronald L. Graham, and David S. Johnson. Some NP-Complete Geometric Problems. In *STOC*, pages 10–22, 1976.

54. M. R. Garey and David S. Johnson. *Computers and Intractability: A Guide to the Theory of NP-Completeness*. W. H. Freeman, 1979.

55. Omer Giménez and Marc Noy. On the Complexity of Computing the Tutte Polynomial of Bicircular Matroids. *Combinatorics, Probability & Computing*, 15(3):385–395, 2006.

56. Robert Goddard. A Method of Reaching Extreme Altitudes. Technical report, Smithsonian Institution, 1919.

57. Carla Gomes, Henry Kautz, Ashish Sabharwal, and Bart Selman. Satisfiability Solvers. In Frank van Harmelen, Vladimir Lifschitz, and Bruce Porter, editors, *Handbook of Knowledge Representation*, pages 89–134. Elsevier, 2008.

58. James Gordon and Liebeck Martin. *Representations and Characters of Groups*. Cambridge University Press, 2001.

59. Ron Graham and Joel Spencer. The Elementary Proof of the Prime Number Theorem. *The Mathematical Intelligencer*, 31(3):18–23, 2009.

60. Ben Green and Terence Tao. The primes contain arbitrarily long arithmetic progressions. *Annals of Mathematics*, 167(1):481–547, 2008.

61. Leonid Gurvits and Warren D. Smith. Definite integration and summation are P-hard, 2008.

62. Dan Gusfield. *Algorithms on Strings, Trees and Sequences: Computer Science and Computational Biology*. Press Syndicate of the University of Cambridge, 1997.

63. J. Havil. *Gamma: exploring Euler's Constant*. Princeton University Press, 2003.

64. Maurice Herlihy and Sergio Rajsbaum. Algebraic Topology and Distributed Computing: A Primer. Technical report, Brown University, 1995.

65. Thomas Hofmeister, Uwe Schöning, Rainer Schuler, and Osamu Watanabe. Randomized Algorithms for 3-SAT. *Theory Comput. Syst.*, 40(3):249–262, 2007.

66. John E. Hopcroft, Wolfgang J. Paul, and Leslie G. Valiant. On Time versus Space and Related Problems. In *FOCS*, pages 57–64, 1975.

67. Ellis Horowitz and Sartaj Sahni. Computing Partitions with Applications to the Knapsack Problem. *J. ACM*, 21(2):277–292, 1974.

68. Philip M. Lewis II, Richard Edwin Stearns, and Juris Hartmanis. Memory bounds for recognition of context-free and context-sensitive languages. In *FOCS*, pages 191–202, 1965.

69. N. Immerman. Nondeterministic space is closed under complement. *SIAM Journal on Computing*, 17:935–938, 1988.

70. Russell Impagliazzo. A Personal View of Average-Case Complexity. In *Structure in Complexity Theory Conference*, pages 134–147, 1995.

71. Russell Impagliazzo, Valentine Kabanets, and Avi Wigderson. New direct-product testers and 2-query PCPs. In *STOC*, pages 131–140, 2009.

72. Jeff Kahan and Kalai Gil. A counterexample to the Borsuk's conjecture. *Bullentin of American Mathematical Society*, 29:60–62, 1193.

73. Ravi Kannan. A probability inequality using typical moments and concentration results. In *FOCS*, pages 211–220, 2009.

74. Ravindran Kannan. Towards Separating Nondeterminism from Determinism. *Mathematical Systems Theory*, 17(1):29–45, 1984.

75. George Karakostas, Richard J. Lipton, and Anastasios Viglas. On the Complexity of Intersecting Finite State Automata. In *IEEE Conference on Computational Complexity*, pages 229–234, 2000.

76. R. M. Karp. Reducibility among combinatorial problems. In R. E. Miller and J. W. Thatcher, editors, *Complexity of Computer Computations*, pages 85–103. Plenum Press, New York, 1975.

77. Richard M. Karp and Richard J. Lipton. Some Connections between Nonuniform and Uniform Complexity Classes. In *STOC*, pages 302–309, 1980.

78. Richard M. Karp and Michael O. Rabin. Efficient Randomized Pattern-Matching Algorithms. *IBM Journal of Research and Development*, 31(2):249–260, 1987.

79. Jonathan Katz and Luca Trevisan. On the efficiency of local decoding procedures for error-correcting codes. In *STOC '00: Proceedings of the thirty-second annual ACM symposium on Theory of computing*, pages 80–86, New York, NY, USA, 2000. ACM.

80. Iordanis Kerenidis and Ronald de Wolf. Exponential lower bound for 2-query locally decodable codes via a quantum argument. *J. Comput. Syst. Sci.*, 69(3):395–420, 2004.

81. Samir Khuller and Yossi Matias. A Simple Randomized Sieve Algorithm for the Closest-Pair Problem. *Inf. Comput.*, 118(1):34–37, 1995.

82. Paul C. Kocher. Timing Attacks on Implementations of Diffie-Hellman, RSA, DSS, and Other Systems. In *CRYPTO*, pages 104–113, 1996.

83. S. Rao Kosaraju. Decidability of Reachability in Vector Addition Systems (Preliminary Version). In *STOC*, pages 267–281, 1982.

84. Dexter Kozen. Lower Bounds for Natural Proof Systems. In *FOCS*, pages 254–266, 1977.

85. Steven Krantz. *Mathematical Apocrypha*. Mathematical Association of America, 2002.

86. S. Y. Kuroda. Classes of Languages and Linear-Bounded Automata. *Information and Control Theory*, pages 207–233, 1964.

87. Peko Lawrence. *Differential Equations and Dynamical Systems*. Springer-Verlag, New York, 1991.

88. H.W. Lenstra, M. S. Manasse, Jr., and J. M. Pollard. The Number Field Sieve. In *STOC*, pages 564–572, 1990.

89. Jérôme Leroux. The General Vector Addition System Reachability Problem by Presburger Inductive Invariants. In *Proceedings of the 2009 24th Annual IEEE Symposium on Logic In Computer Science*, pages 4–13. IEEE Press, 2009.

90. Leonid Levin. Universal Search Problems. *Problems of Information Transmission*, 9(3):265–266, 1973.

91. Leonid A. Levin. Problems, Complete in "Average" Instance. In *STOC*, page 465, 1984.

92. Richard Lipton and Anastasios Viglas. On the Complexity of SAT. In *FOCS*, pages 459–, 1999.

93. Richard J. Lipton. Polynomials with 0—1 Coefficients that Are Hard to Evaluate. In *FOCS*, pages 6–10, 1975.

94. Richard J. Lipton. A Necessary and Sufficient Condition for the Existence of Hoare Logics. In *FOCS*, pages 1–6, 1977.

95. Richard J. Lipton. Straight-line complexity and integer factorization. In *Algorithmic Number Theory Symposium*, pages 71–79, 1994.

96. Richard J. Lipton and Daniel P. Lopresti. Delta Transformations to Simplify VLSI Processor Arrays for Serial Dynamic Programming. In *ICPP*, pages 917–920, 1986.

97. Richard J. Lipton and Jeffrey F. Naughton. Clocked Adversaries for Hashing. *Algorithmica*, 9(3):239–252, 1993.

98. Richard J. Lipton, Arnold L. Rosenberg, and Andrew Chi-Chih Yao. External Hashing Schemes for Collections of Data Structures. *J. ACM*, 27(1):81–95, 1980.

99. Richard J. Lipton and Robert Sedgewick. Lower Bounds for VLSI. In *STOC*, pages 300–307, 1981.

100. Richard J. Lipton, Larry Synder, and Yechezkel Zalcstein. Isomorphism Problems for Finite Groups. In *John Hopksins Conference*, 1976.

101. Richard J. Lipton and Yechezkel Zalcstein. Word Problems Solvable in Logspace. *J. ACM*, 24(3):522–526, 1977.

102. Eugene M. Luks. Isomorphism of Graphs of Bounded Valence can be Tested in Polynomial Time. *J. Comput. Syst. Sci.*, 25(1):42–65, 1982.

103. Yuri Matiyasevich. *Hilbert's Tenth Problem*. MIT Press, 1993.

104. Ernst W. Mayr. An Algorithm for the General Petri Net Reachability Problem. In *STOC*, pages 238–246, 1981.

105. Gary L. Miller. Riemann's Hypothesis and Tests for Primality. *J. Comput. Syst. Sci.*, 13(3):300–317, 1976.

106. Raymond E. Miller and James W. Thatcher, editors. *Complexity of computer computations*. Plenum Press, 1972.

107. Marvin Minsky. *Computation: Finite and Infinite Machines*. Prentice-Hall, 1967.

108. Joseph S. B. Mitchell. A PTAS for TSP with neighborhoods among fat regions in the plane. In *SODA*, pages 11–18, 2007.

109. Kary Mullis. *Dancing Naked in the Mind Field*. Random House, 2000.
110. Eugene W. Myers. Toward Simplifying and Accurately Formulating Fragment Assembly. *Journal of Computational Biology*, 2(2):275–290, 1995.
111. Ashwin Nayak. Optimal Lower Bounds for Quantum Automata and Random Access Codes. In *In Proceedings of 40th IEEE FOCS*, pages 369–376, 1999.
112. Noam Nisan. Psuedorandom Generators for Space-Bounded Computation. In *STOC*, pages 204–212, 1990.
113. Andrew Odlyzko. The 10 to 22nd zero of the Riemann zeta function. *Amer. Math. Soc., Contemporary Math*, 290(1):139–144, 2001.
114. George Polyá. *How to Solve It*. Princeton University Press, 1957.
115. MIchael Rabin. Probabilistic algorithm for testing primality. *Journal of Number Theory*, 12(1):128–138, 1980.
116. Ran Raz. A Parallel Repetition Theorem. *SIAM J. Comput.*, 27(3):763–803, 1998.
117. Alexander A. Razborov and Steven Rudich. Natural Proofs. *Journal of Computer and System Sciences*, 55(1):24–35, August 1997.
118. John E. Savage. Computational Work and Time on Finite Machines. *J. ACM*, 19(4):660–674, 1972.
119. W. J. Savitch. Relational between nondeterministic and deterministic tape complexity. *Journal of Computer and System Sciences*, 4:177–192, 1980.
120. Lowell Schoenfeld. Sharper Bounds for the Chebyshev functions. *Mathematics of Computation*, 30(134):37–360, 1976.
121. Adi Shamir. IP=PSPACE. In *FOCS*, pages 11–15, 1990.
122. N. J. Sloane and Aaron Wyner. *Claude Shannon, Collected Papers*. IEEE Press, 2006.
123. Roman Smolensky. Algebraic Methods in the Theory of Lower Bounds for Boolean Circuit Complexity. In *STOC*, pages 77–82, 1987.
124. Daniel A. Spielman and Shang-Hua Teng. Smoothed analysis of algorithms: why the simplex algorithm usually takes polynomial time. In *STOC*, pages 296–305, 2001.
125. Larry J. Stockmeyer. On Approximation Algorithms for #P. *SIAM J. Comput.*, 14(4):849–861, 1985.
126. Robert Szelepcsényi. The Method of Forced Enumeration for Nondeterministic Automata. *Acta Inf.*, 26(3):279–284, 1988.
127. George Szpiro. *Poincare's Prize*. Dutton, 2007.
128. Terence Tao. *Structure and Randomness: Pages from Year One of a Mathematical Blog*. AMS, 2009.
129. Leslie G. Valiant. Holographic Algorithms (Extended Abstract). In *FOCS*, pages 306–315, 2004.
130. J. H. van Lint. The van der Waerden conjecture: two proofs in one year. *The Mathematical Intelligencer*, 4(2):72–77, 1982.
131. R. A. Wagner and M. J. Fischer. The string-to-string correction problem. *J. ACM*, 21(1):168–173, 1974.
132. Andrew Wiles. Modular elliptic curves and Fermat's last theorem. *Annals of Mathematics*, 143(3):443–572, 1995.
133. Ryan Williams. Matrix-vector multiplication in sub-quadratic time: (some preprocessing required). In *SODA*, pages 995–1001, 2007.
134. Ryan Williams. Time-Space Tradeoffs for Counting NP Solutions Modulo Integers. In *IEEE Conference on Computational Complexity*, pages 70–82, 2007.
135. Ryan Williams. Alternation-Trading Proofs, Linear Programming, and Lower Bounds. In *STACS*, pages 669–680, 2010.
136. Philipp Woelfel. Bounds on the OBDD-size of integer multiplication via universal hashing. *J. Comput. Syst. Sci.*, 71(4):520–534, 2005.
137. Jin yi Cai, Xi Chen, and Pinyan Lu. Graph Homomorphisms with Complex Values: A Dichotomy Theorem. *CoRR*, abs/0903.4728, 2009.
138. Alexander Zarelua. On congruences for the traces of powers of some matrices. *Proceedings of the Steklov Institute of Mathematics*, 263(1):78–98, 2008.
139. David Zuckerman. Linear degree extractors and the inapproximability of max clique and chromatic number. In *STOC*, pages 681–690, 2006.

Index

List of Names

A

Aaronson, Scott 36, 192
Aceto, Luca 98
Adjan, Sergei 221
Adleman, Len 28, 150, 202
Aldrin, Buzz 62
Allender, Eric 112
Alon, Noga 190
Anand, Bhupinder 21
Andoni, Alexandr 181
Appel, Kenneth 223
Armstrong, Neil 62
Arnold, Vladimir 166
Arora, Sanjeev 44, 222
Arrow, Kenneth 224
Ax, James 211

B

Babai, László 61, 217
Bacon, Dave 21
Barak, Boaz 36, 44, 61
Bardeen, John 196
Barnum, Howard 21
Barrington, David 85, 88, 119, 187, 212
Beame, Paul 21, 36, 59, 60, 225
Bellman, Richard 179
Bender, Edward 212
Bennett, Charles 74
Bhattacharyya, Paul 225
Birkhoff, George 223
Black, Nick 183
Blum, Avrim 9
Blum, Lenore 9

Blum, Manny 9
Bombieri, Enrico 14
Boneh, Dan 67, 74, 143
Borodin, Allan 186
Borsuk, Karol 219
Brassard, Giles 74
Brattain, Walter 196
Brent, Richard 86
Britton, John 222
Brown, Gordon 71
Brumley, David 143
Bryant, Randy 157
Burnside, William 88, 221
Buss, Sam 86
Byrne, Drew 17

C

Cai, Jin-Yi 119, 197, 211
Camarena, Omar 114
Cardoza, E. 97
Carroll, Lewis 56
Carroll, Ted 218
Chakrabarti, Amit 75
Chandra, Ashok 49, 86, 185
Chapman, Noyes 30
Chen, Xi 122
Chen, Zhi-Zhong 113
Chesnutt, Charles 73
Chomsky, Noam 136
Chow, Timothy 31, 48, 65, 114
Churchill, Winston 72
Ciobac, Stefan 188
Ciobaca, Stefan 21
Clarke, Arthur 190